KU-514-705

Bright Sparks

AMAZING DISCOVERIES, INVENTIONS & DESIGNS BY WOMEN

OWEN O'DOHERTY

THE O'BRIEN PRESS
DUBLIN

First published 2018 by
The O'Brien Press Ltd,
12 Terenure Road East, Rathgar,
Dublin 6, D06 HD27, Ireland.
Tel: +353 1 4923333; Fax: +353 1 4922777
E-mail: books@obrien.ie
Website: www.obrien.ie
The O'Brien Press is a member of Publishing Ireland.

ISBN: 978-1-78849-054-2

10 9 8 7 6 5 4 3 2 1
22 21 20 19 18

Printed by L&C Printing Group, Poland.
The paper in this book is produced using pulp from managed forests.

Acknowledgements

I am grateful to many people for their help, support and hard work in turning this idea into a finished book. For crucial encouragement and advice at the early stages I have to thank Jesse Jones, Bill Christian, Cormac Kinsella, Sallyanne Godson, Toby Scott, Geraldine O'Doherty, Bini Godson and in particular Jennie Flynn for her help and direction. Thanks to Roland Bosbach and Dennis McNulty among others for suggesting inventors and making the early research such fun.

While there are many available sources for the invention stories, Deborah Jaffé's *Ingenious Women* has to be mentioned as an inspiring introduction to the history of female inventors.

The illustration on p.22 is reproduced from a photo with permission from the Irish Defence Forces.

Thanks to all at O'Brien Press who helped develop the proposal through to completion, from the first talks with Michael O'Brien to the later work with Eoin O'Brien, Emma Byrne and everyone else. Editor Síne Quinn was the critical guide through the writing process and I am really grateful for her expertise, eye for detail and ability to calmly deal with a total novice.

Several people kindly commented on draft material and offered advice during the writing. Thanks to Toby, Robyn and Hadley for their incisive comments over a wonderful lunch in Sligo. Thanks also to Tim Gill, Tara Whelan, Liam Donnelly and Catherine Godson for really useful comments and advice. Special thanks to my father Michael for continual suggestions, queries and encouragement throughout.

Finally, thanks to James and Anna for the conversation that started the whole idea in the first place and for the love and conversations that we have all the time, and of course to my wife Lisa for everything.

Dedication

This book is for Anna and James,
may all your journeys be inventive,
and for my mother Moya, a
champion of ingenuity.

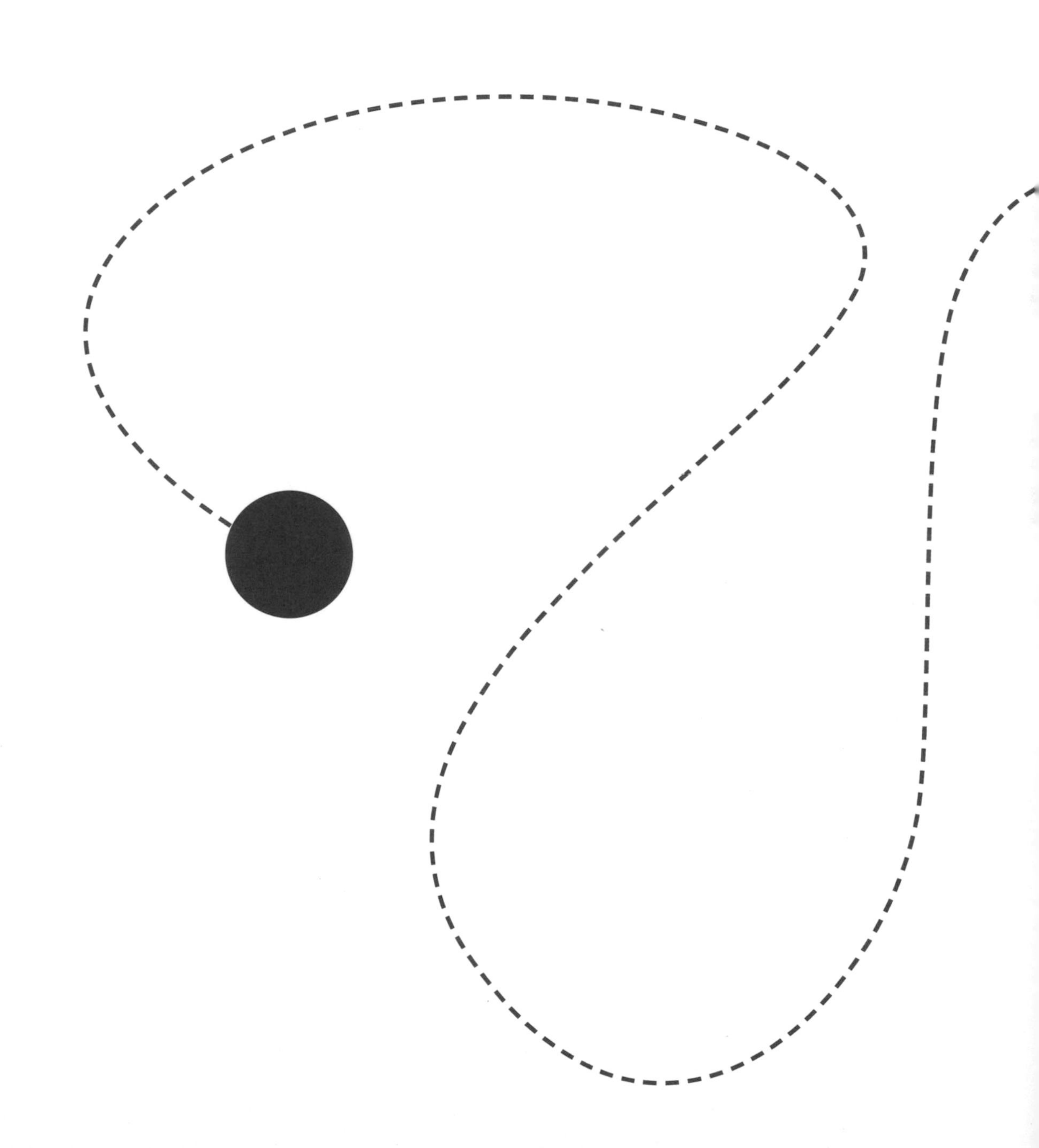

Hi!

This book is about discoveries, inventions and designs that have changed the lives of people all over the world, and the remarkable women who devised them.

There are various ways these innovations have changed our world. Some have saved lives; others are simple everyday tools. There are innovations that have changed how we see the world and others that are making changes we will see in the future.

Women everywhere, from different times and every type of background, have made discoveries and created inventions in all kinds of situations. There are those who had to overcome great challenges or prejudice. A number of them won notable awards for their achievements while others were overlooked and unrecognised or had to pass their ideas on. And they invented at all ages: two of the inventors featured were only twelve and thirteen when they came up with their ingenious ideas.

These scientists and inventors were inspired to create their innovations for many reasons: witnessing an accident, being fascinated by the hidden mysteries of deep space or microscopic organisms, or just seeing something that they thought they could improve. Whatever their reasons, they have all had certain things in common: being curious about how the world works, wanting to change things and working hard to explore their ideas.

While you might already know about many of the discoveries and inventions here, you might not have heard of the people who did them. Often in the past when women invented things their stories were not told, or not told properly or not told as loudly as stories of inventions by men. But there have been many impressive women inventors, scientists and designers – in fact, far too many to fit into a book this size. If you know of an inventor that you think more people should hear about, then why not find out more yourself, and share their story with the world?

But as well as reading about these important achievements, you'll also find tips to help you with your own projects: how to look at things differently, see how you can make a change, try out your ideas and then make your invention. If you have an idea for a new way of doing things or you notice something new about the world, see if you can explore it further. You never know – you just might be the first!

Contents

Innovation

Discoveries, inventions and designs are all forms of innovation

Innovation means making changes and we are continually changing the world around us with new knowledge, devices, materials or ways of doing things.

To put things simply, we can talk about the things that have changed our world in three rough groups: discoveries, inventions and designs.

M. E. KNIGHT.
Paper-Bag Machine.
No. 220,925. Patented Oct. 28, 1879.

Investigating diseases and finding new medicines

Exploring outer space

DISCOVERIES

Things that already existed (usually before we did) but we didn't know about them.

Learning about the past through fossils or archaeology

Uncovering how the Earth is formed

Working out the laws of physics that our universe follows

Follow this path through the book to see stories of amazing discoverers, inventors and designers!

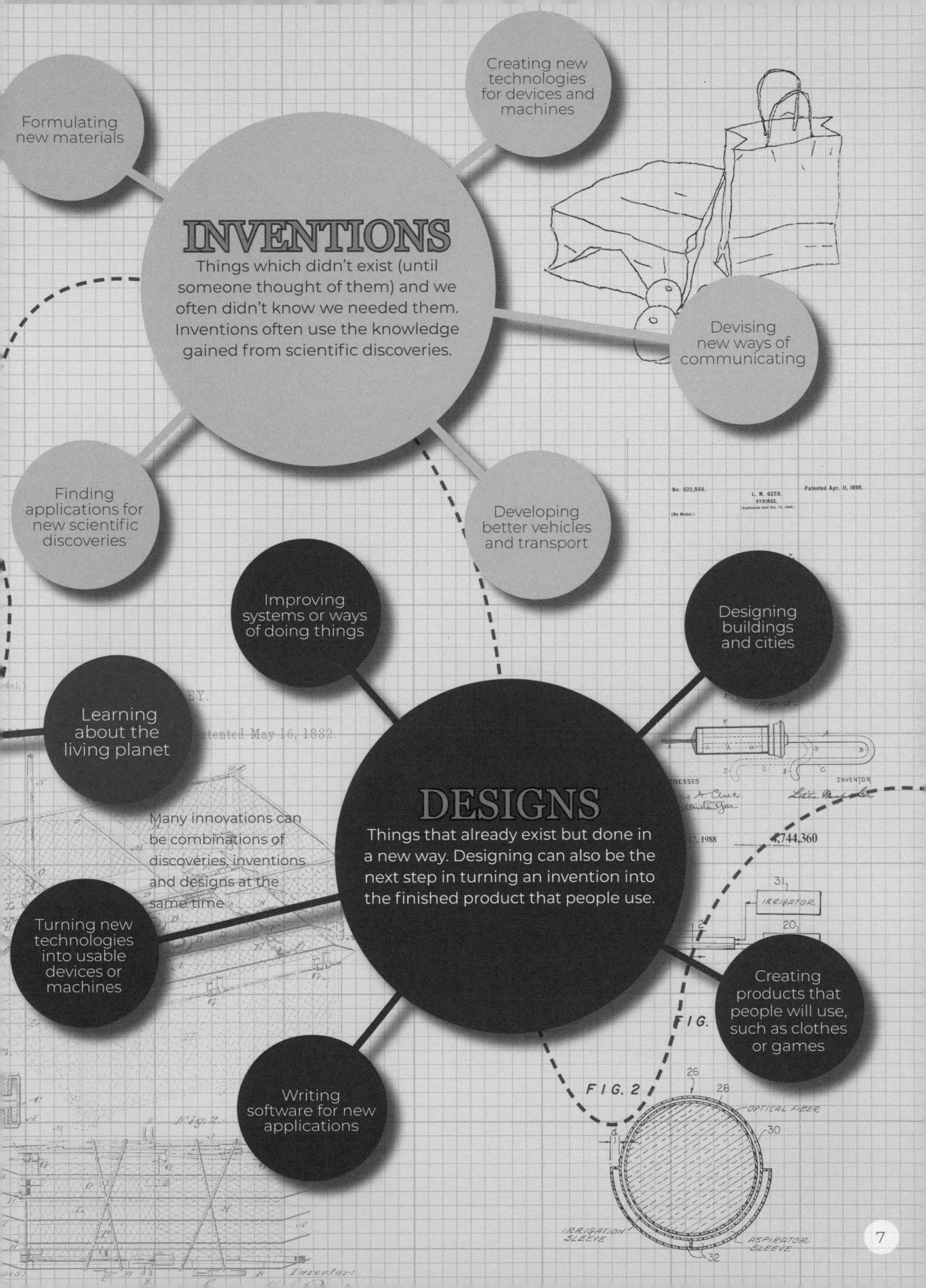
Formulating new materials
Creating new technologies for devices and machines
INVENTIONS
Things which didn't exist (until someone thought of them) and we often didn't know we needed them. Inventions often use the knowledge gained from scientific discoveries.
Devising new ways of communicating
Finding applications for new scientific discoveries
Developing better vehicles and transport
Improving systems or ways of doing things
Designing buildings and cities
Learning about the living planet
Many innovations can be combinations of discoveries, inventions and designs at the same time
DESIGNS
Things that already exist but done in a new way. Designing can also be the next step in turning an invention into the finished product that people use.
Turning new technologies into usable devices or machines
Creating products that people will use, such as clothes or games
Writing software for new applications
No. 622,848.
L. M. GEER.
SYRINGE.
Patented Apr. 11, 1899.
(No Model.)
Patented May 16, 1882.
INVENTOR
4,744,360
IRRIGATOR
FIG. 2
OPTICAL FIBER
IRRIGATION SLEEVE
ASPIRATOR SLEEVE

1. APGAR score

The system for checking the health of newborn babies

When a baby is born, doctors and midwives have to decide quickly if they are healthy or need extra care. All babies are different, and the Apgar Score is a system that makes checking babies' health much quicker and easier. Since its invention by Dr Virginia Apgar it has helped save the lives of millions of babies. Now the doctors can treat babies in a much faster and more efficient way.

A score from zero to two is given for each of five things: skin colour, pulse, reflexes, activity and breathing. The scores are added up and the result is used to help decide quickly if the baby needs treatment.

As a play on Apgar's name the acronym APGAR was later given to the scores standing for Appearance, Pulse, Grimace, Activity and Respiration.

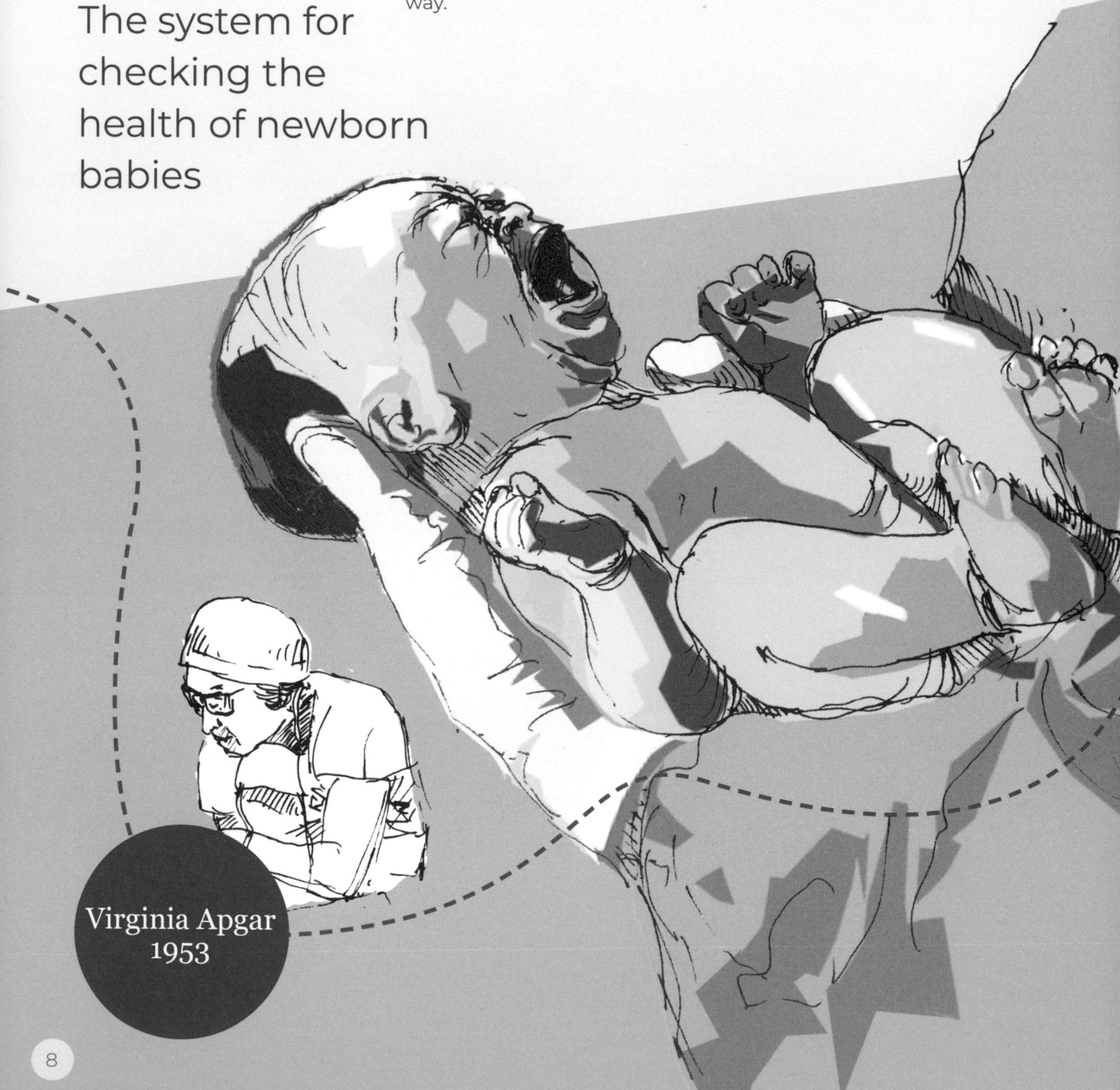

Virginia Apgar
1953

2. Mid-Atlantic Ridge

The discovery that proved continental drift

Marie Tharp was a geologist who calculated the shape of the Atlantic Ocean's sea bed by using information from mapping expeditions. Her calculations showed that there was a large rift running down the middle of the ocean: the Mid-Atlantic Ridge.

At that time some scientists had suggested that the continents move very slowly over time, rather than always staying in fixed positions – a process called 'continental drift' – but not many people believed this theory. The Mid-Atlantic Ridge is the edge between two moving continental plates: America and Eurasia. Tharp's discovery proved that continental drift was real and that the continents move and change shape over time.

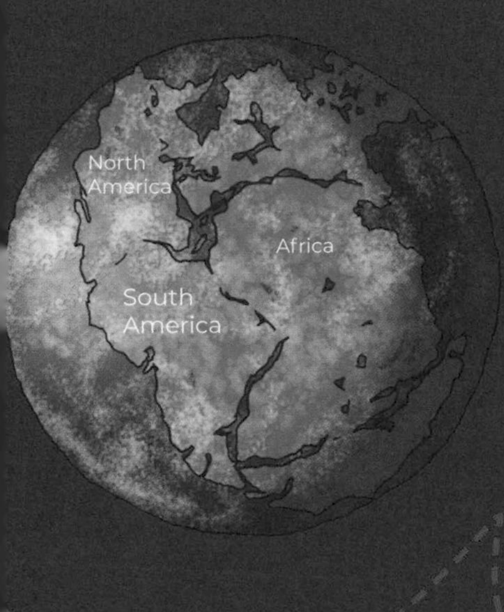

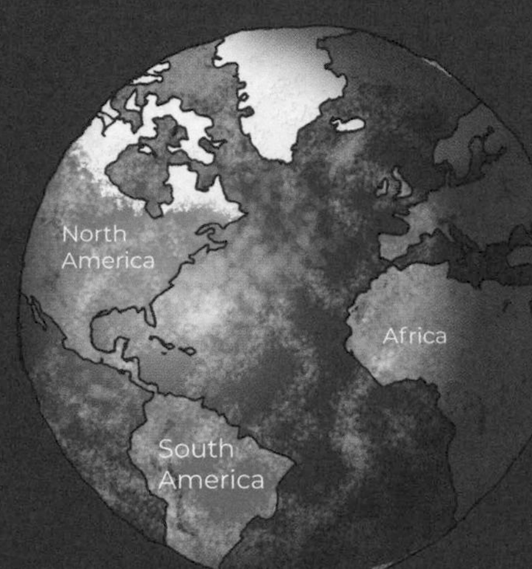

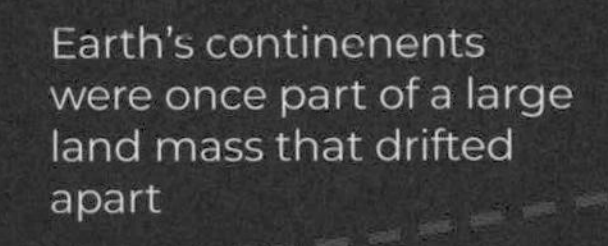
Earth's continenents were once part of a large land mass that drifted apart

Marie Tharp
1952

3. Retractable dog lead

If you've ever been dragged along by an energetic dog, you'll know how useful retractable dog leads can be. The first one was patented by Mary Delaney 110 years ago. The lead changes length, allowing a dog to roam more freely on walks while still keeping them under control. The lead is coiled or rolled on a spring-loaded drum that can be pulled and made longer as the dog moves away from its walker, but at the press of a button the spring can pull the lead back to make it shorter again.

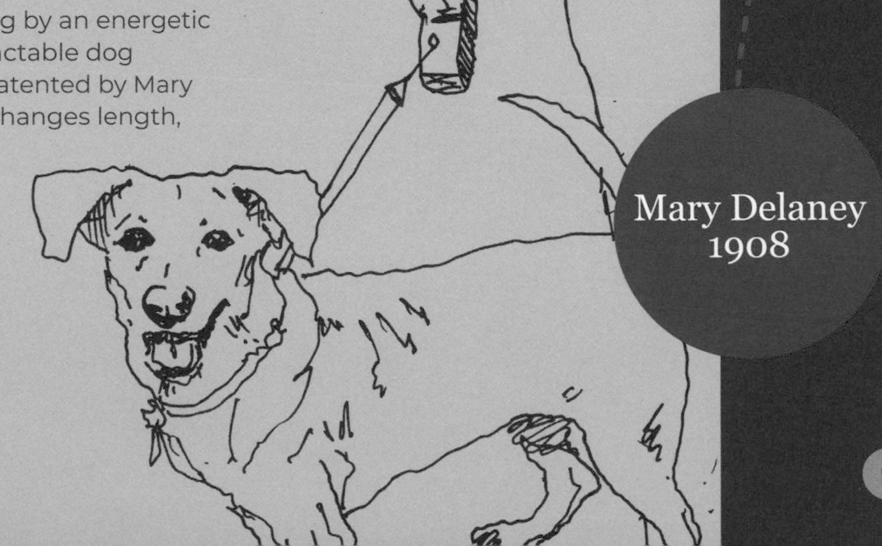

Mary Delaney
1908

4. Suspension bridge

A bridge design for spanning wide rivers

Engineering owes an innovation to Sarah Guppy, who patented a way of building a suspension bridge that could cross a large river. She had seen this problem in Bristol in England. A suspension bridge is one where a road is hung or suspended from cables above it. Suspension bridges can span much greater distances than traditional arched bridges.

When Guppy patented her invention in 1811 women did not usually work in engineering. In fact, Guppy passed on her patent to engineers Thomas Telford and Isambard Kingdom Brunel, because she didn't think it was appropriate for her to be working as an inventor. Although she didn't design the final Clifton Bridge, Guppy's ideas can still be seen in the finished structure.

Guppy's other inventions included a design for a coffee urn that could also keep toast warm and a bed with bars that doubled as a gym for women to exercise in the privacy of their bedrooms.

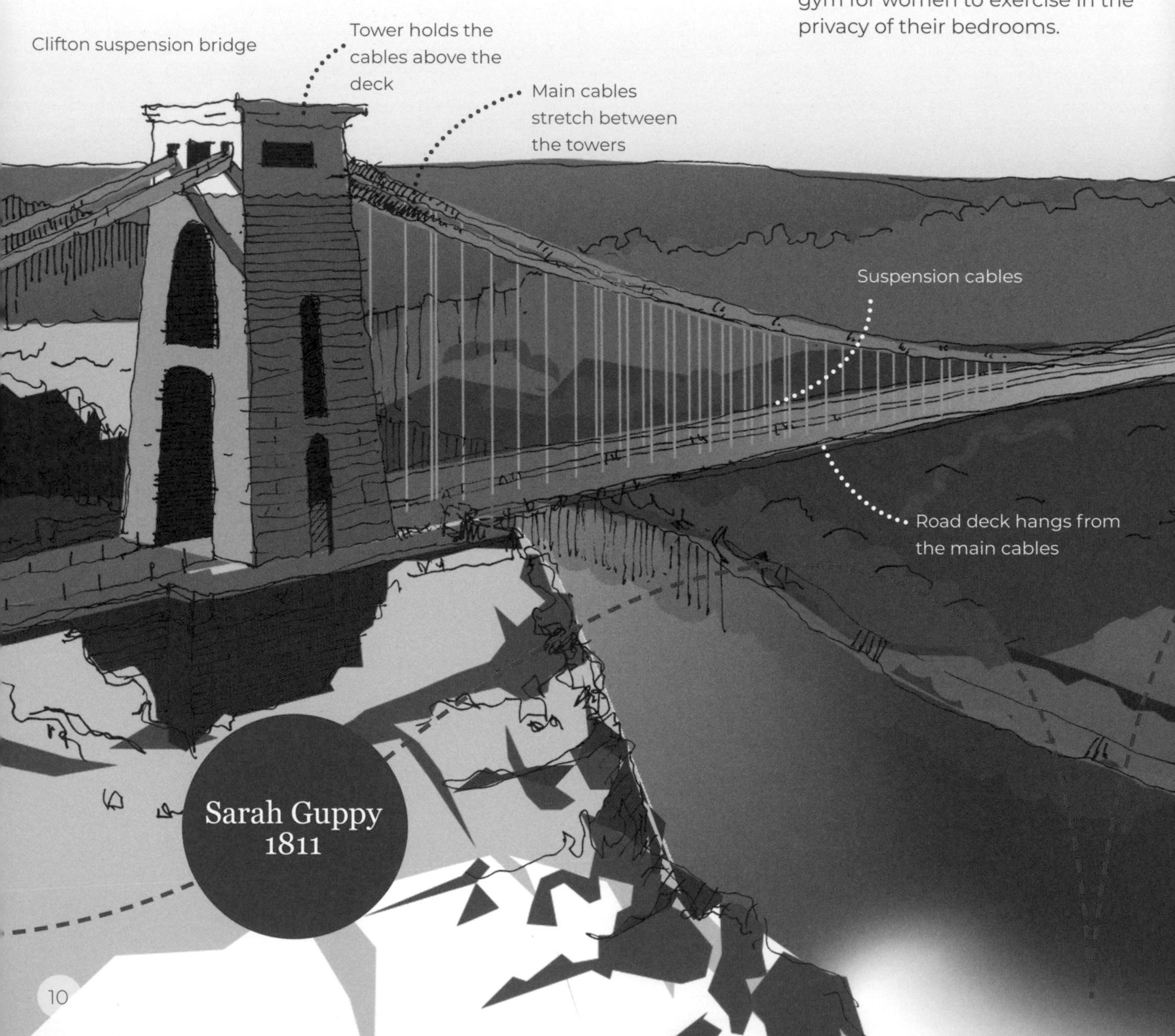

5. Pedal bin

An ingenious everyday invention

Lillian Gilbreth
1920s

The pedal bin was invented by Dr Lillian Gilbreth, a psychologist and industrial designer. Gilbreth specialised in ergonomics, which is the science of how to design things so they are safe and easy to use, and she was an expert on helping people to work efficiently. Her ingenious invention meant that people could open the bin by pressing the pedal with their feet, leaving both hands free to hold things. Pedal bins are used in many situations where hands-free use is helpful, from kitchens to hospitals.

Gilbreth was a remarkable person – a successful inventor, scientist and the mother of twelve children. Her other ergonomic inventions included putting shelves to store eggs and butter inside the door of refrigerators.

Tower built on top of pier

6. Knapsack parachute

Käthe Paulus
1890s

Käthe Paulus was a German aerial acrobat who was famous for entertaining crowds with her stunts and daring parachute jumps from hot air balloons and aircraft.

Because Paulus was an early pioneer of parachuting, she had to invent a lot of the equipment herself. She is credited with inventing the folding parachute (knapsack parachute) in the 1890s. Before this, parachutes were cumbersome objects that had to be carried by the person jumping. Paulus was the first German woman to jump out of a plane, making more than 165 parachute jumps in her lifetime.

Paulus jumping from a balloon

7. Coding

Writing the languages used by computers

Every laptop, tablet, smartphone or desktop computer uses computer programs (called software) to operate.

We often think of computers as modern devices, so it might come as a surprise to know that today's computer programs have their roots in an invention made by Ada Lovelace in 1843 – 175 years ago!

Her friend, Charles Babbage, had invented a mechanical calculator called the Analytical Engine, an early computer. Babbage had devised the engine as a way to do calculations that we'd now consider simple. Lovelace studied how the engine worked and realised that it could be used to carry out more than basic calculations.

She wrote the very first computer algorithm, or a set of steps used to solve a problem. These could enable the engine to carry much more complicated tasks. For this invention Lovelace has been called the first computer programmer or coder.

8. Structure of DNA

Seeing the DNA molecules inside every living cell

Rosalind Franklin 1952

DNA is a type of molecule found in every cell of our bodies and it is the part that tells each cell what to be; for example: a bone, brain or skin cell. Chemist and scientist Rosalind Franklin made visible the beautiful and complicated structure of the DNA molecule.

Franklin used X-ray crystallography to make pioneering photographs of DNA. Her incredible photos revealed the intricate structure of DNA: two spirals wrapped around each other, called a double-helix. One of her photos, Photograph 51, took 100 hours of X-ray exposure and was later used to work out the famous double-helix model of DNA.

Her groundbreaking research helped unlock the problem of how DNA works. Franklin died young and was not credited when the Nobel Prize was later awarded to two other scientists for the work that her discovery had helped.

'Base pairs' of nucleotides

Two spiral-shaped backbones made up of phosphate and deoxyribose holding the base pairs together

9. Coffee filter

Melitta Bentz 1908

Making a good cup of coffee can be harder than you think. Melitta Bentz decided to tackle the problem of the gritty and bitter-tasting coffee grounds that lingered at the bottom of coffee cups. She finally solved it and patented the first paper coffee filter in 1908. The filter consisted of a paper cone through which the coffee is poured, straining out the grounds. The company she founded, Melitta, still makes coffee filters to this day.

10. Syringe

Letitia Geer
1899

The modern needles for injecting medicine

Needle point

Cartridge holding the medicine

Finger grips that allow someone to use the syringe with only one hand

Plunger to push the medicine out through the needle

Needles used to give injections are called syringes. While people have been using needles for medicine for hundreds of years, the whole process was made much easier when Letitia Geer invented the modern syringe.

Geer's clever idea was to design a needle which included grips so that a user could hold and operate the needle in one hand, leaving their other hand free to use with the patient.

11. Aquarium

Jeanne Villepreux-Power
1832

Seeing sea creatures up close

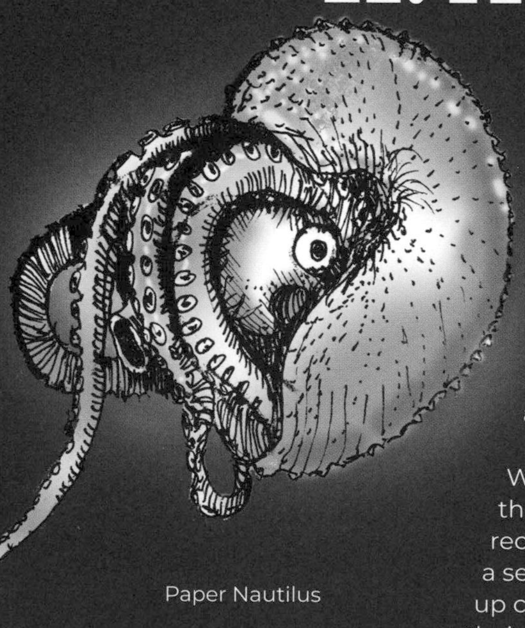

Paper Nautilus

Jeanne Villepreux-Power invented the first aquarium, although her invention wasn't given this name until later. Villepreux-Power was a dressmaker and designer in Paris at a young age, designing for wealthy clients. She later moved to Sicily. For over 20 years she carried out extensive work studying and recording the lives of underwater plants and animals in the sea around the island.

While she wasn't a trained scientist, she studied the lives of the creatures in great detail and followed scientific methods in recording what she learned. To explore the life of the nautilus, a sea creature similar to an octopus, she needed to see them up close over a long time to learn more about how they grow their shells. She devised a glass case to hold sea water that the creatures could live in while she studied them. This is the first known aquarium.

12. Nuclear fission

Understanding how atoms split

Smaller krypton nucleus

Neutron added to nucleus

Energy release

Neutrons released

Large uranium nucleus

Smaller barium nucleus

Nuclear fission is the process where the nucleus of an atom is split into smaller nuclei while releasing neutrons and energy. Physicist Lise Meitner first explained and named this process.

In the 1930s scientists were discovering new, radioactive elements with larger nuclei and were racing to discover more. Meitner worked with a chemist, Otto Hahn, to experiment with creating new elements. They bombarded the nuclei of the radioactive element uranium with neutrons. Hahn carried out the experiments while Meitner worked to understand and explain the process.

They discovered that instead of creating a larger atom, they were creating smaller ones. Meitner worked out that the uranium nuclei were splitting and breaking down into smaller elements while also releasing more neutrons and energy. They had discovered nuclear fission, the powerful energy source that drives nuclear power stations.

While Meitner and Hahn had worked together, only Hahn was awarded a Nobel Prize. The award jury did not understand the importance of Meitner's role in the discovery, and she was overlooked. However, she later received other awards and the element meitnerium was named in her honour.

The inventor's glossary

Application
An application is a use that technology or scientific research is put to. Design is a method of turning research or technology into applications.

Biology
The study of living things.

Chemistry
The study of the substances that make up matter and how they react with energy and each other.

Coding
Writing computer programs using a computer language. Also called computer programming.

Data
Information, such as facts, used to help make decisions. For example, the results of scientific experiments are data.

Designing
The process of taking an idea, imagining the finished product and working out how to make it. Buildings are an example of things that have been designed.

Design brief
A description of what a design or invention is meant to achieve. It is useful for guiding decisions and for checking later on that a design achieves what it was supposed to.

Device
A piece of technology or equipment made for a particular purpose.

Engineer
Someone who designs technology such as machines, structures or systems. Engineers often work on turning inventions or scientific research into working devices or systems.

Knight's (no. 33) patent drawing describing her paper-bag-making invention

Lovelace's (no. 7) diagram explaining her idea for the first computer algorithm

Entrepreneur
Someone who decides to set up a business and run it.

Ergonomics
The science of how to make devices or methods useful, safe and easy to use.

Ethics
Beliefs about whether behaviour is right or wrong. For inventions, as well as considering whether a product works, an inventor needs to consider if it would be harmful to people or the environment. Some inventions can be controversial, with people being divided about whether they are ethically right or not.

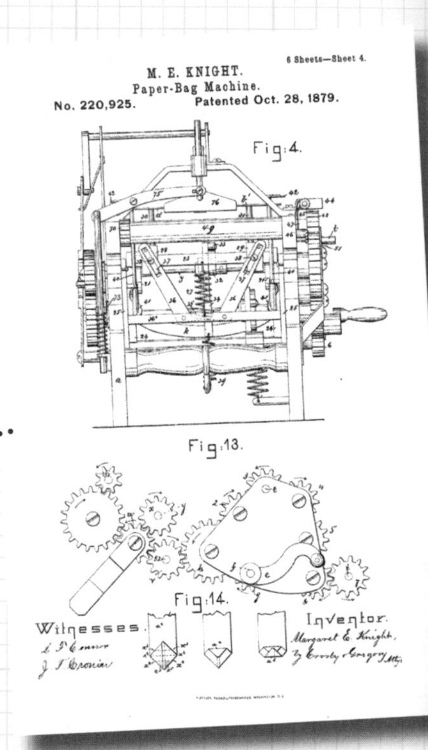

Experiment
A type of research that uses controlled investigations to find out new information. Scientists use experiments to test whether a hypothesis is true or not.

Genetics
The science of how features pass from parents to their young through genes, which are made up of DNA.

Hypothesis
A starting idea or explanation for something based on incomplete information. A hypothesis can then be tested for accuracy through experiments to find out whether it is true or not.

Infographics
Illustrations or diagrams that explain data to make it easier to understand and use.

Innovation
A new and improved way of doing something. Inventions, discoveries and designs are all different ways of innovating.

Intellectual property
Intellectual property (or IP) is anything that is created by human intellect or mind, including ideas, inventions, discoveries, designs, music or art. IP can be valuable, so inventors can protect their IP from being stolen or copied with patents that state they are the owner. Designers can protect their IP by declaring it is copyright.

Invention
The creation of a new device, system or product.

Mass production
Making objects or products in very large numbers. Mass production makes products more available by using systems to make them more cheaply.

Observation
Taking in information from things around you.

Physics
The study of matter and physical objects and how they react with energy and forces.

Product
Anything that is made and that people need or find useful or attractive.

Prototype
An early version of a product or system that is made during the design process to test whether it works and to research any problems. Prototypes are important because discovering a problem with a prototype is less costly than discovering a problem after an invention has been mass produced.

Research
The process of investigating and finding things out to add to your knowledge.

Software
The programs that tell computers what to do and how to do it. They are written in computer languages that the machines can read and use.

Technology
Technology means the useful devices or systems that are developed from scientific knowledge, for example, engines, medical devices or computer software.

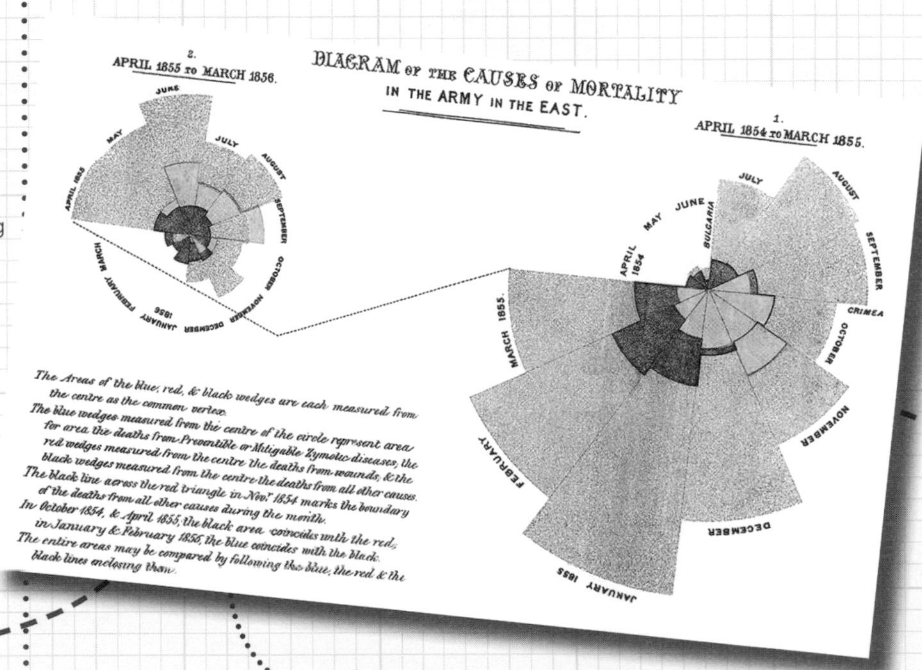

This diagram was made by Nightingale (no. 44) to help people understand her data measuring the different causes of deaths among patients. The sizes of the wedge shapes represent the numbers of deaths from different causes in each month

STEM
Stands for Science, Technology, Engineering and Maths, the school subjects that together help increase people's knowledge of technology and scientific research.

System
An organised way of doing things, where different activities or methods are connected together to give the desired result. Designing or improving systems is a way of innovating.

Theory
A theory is an idea or explanation of how things work based on the information available. Theories are flexible and can change as new knowledge or research becomes available.

User
The person who will use a device, system or product. Designers research the needs of the user in order to help make decisions during the design process.

13. Modern kitchen

A revolutionary design for cooking

Some designs improve things by reorganising them so that they work better together. Modernist architect Grete Schütte-Lihotsky's reorganised kitchen design aimed to reduce the amount of work for users.

Using ergonomic techniques, Schütte-Lihotsky measured the activities and movements of cooking and cleaning, and designed equipment to suit. Her designs were used in 10,000 workers' homes in Frankfurt. Her mass-produced work changed kitchen design.

She had always wanted to use her skills for improving the conditions of working people, and later designed new cities being built for workers in the Soviet Union. During the Second World War she was captured by Nazi secret police in Vienna when she was helping the resistance. Freed by Allied troops at the end of the war, Schütte-Lihotsky lived to 103 years old.

Grete Schütte-Lihotsky
1926

14. Mouldable rubber

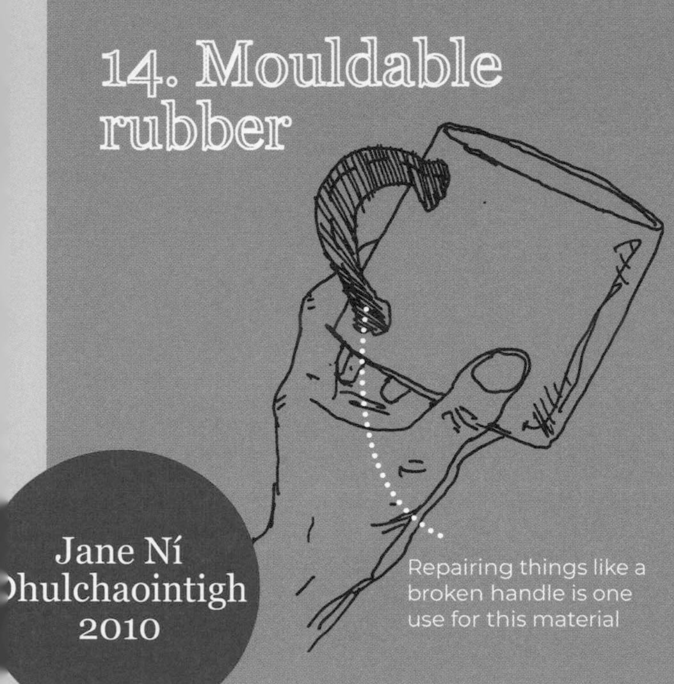

Repairing things like a broken handle is one use for this material

Jane Ní Dhulchaointigh
2010

Irish designer Jane Ní Dhulchaointigh became a scientific expert in order to develop her invention: Sugru. This is a soft material that can stick to many surfaces. It can be shaped with your fingers like modelling clay. But when it sets it becomes a more solid rubber material. It can be used for many applications such as repairing things, adding features like grips or protective pads to equipment, or making whole new parts.

Ní Dhulchaointigh's idea was for a material that would allow people to repair or improve things they own rather than buy new things. While she had a clear idea of what she wanted her invention to do, the material she wanted didn't exist. She worked for a long time with scientists researching and developing her invention, learning about the science of materials in order to turn her idea into a reality.

15. Boom microphone

Dorothy Arzner was an innovative film director in Hollywood who invented an ingenious solution to a new problem faced by the film industry. The problem was how to record actors moving around a film set for the new 'talking pictures' without having a microphone appear in the picture. This problem hadn't existed with silent movies.

How sound is recorded for films

Arzner's innovation, now known as the boom microphone, allowed sound recordists to keep the microphone close enough to pick up actors' voices without being seen by the cameras filming the action.

The device consists of a rod (or boom) which holds a microphone close to a speaker to capture the sound. The sound device is still used today in film to record actors' voices. Arzner didn't patent her invention, which was patented by another inventor a year later.

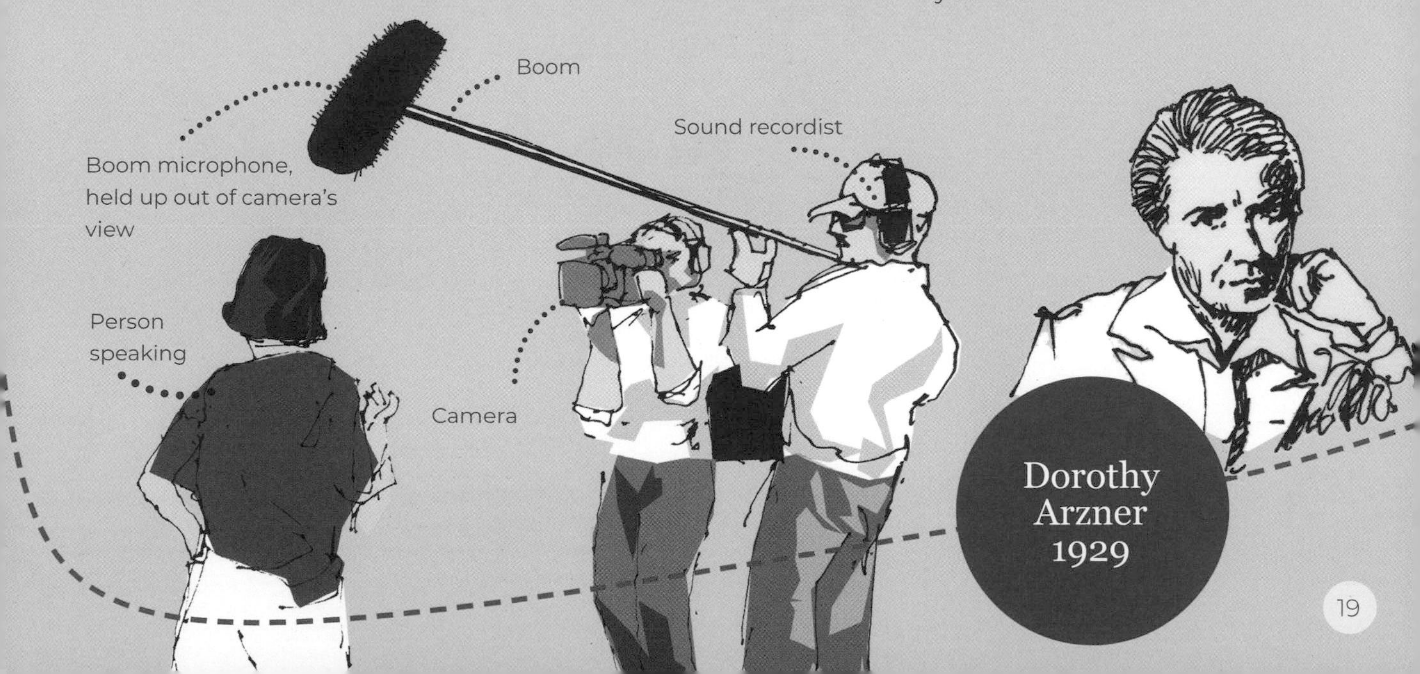

Dorothy Arzner
1929

16. Boosting seeds

Ciara Judge, Emer Hickey, Sophie Healy-Thow 2013

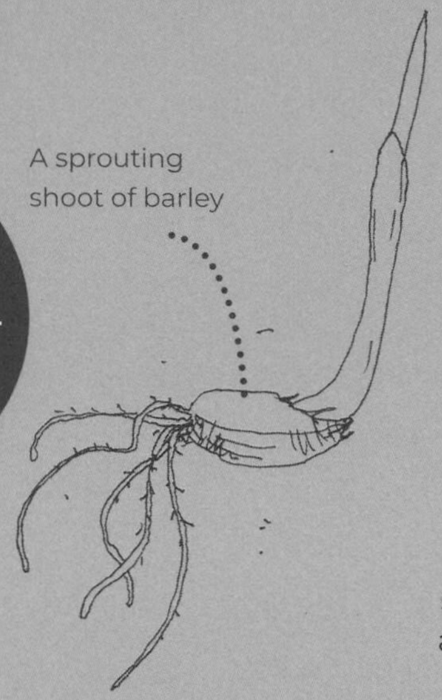

A sprouting shoot of barley

There are young new scientists and inventors at work all the time. In 2013 three secondary school students, Ciara Judge, Emer Hickey and Sophie Healy-Thow, discovered a way to speed up germination. They were curious about the way a kind of bacteria that lives on the roots of vegetables (called rhizobium) might affect germination. They experimented, measuring the germination of 11,000 seeds (that's a LOT of seeds). They recorded all their results so that they could study which combinations of seeds and bacteria gave the best growth.

They discovered that the bacteria could make wheat, barley and oat seeds germinate twice as quickly. They also found that using this method produced 74% more barley from the same amount of seed. Their discovery means that much more food can be produced in a shorter time. This could help reduce global hunger and also means less chemical fertilisers are used.

17. Coade stone

A hard wearing stone for buildings

A material called Coade stone became well known in the late 1700s and 1800s for its durability. It was used for making statues and decorations on buildings. Eleanor Coade, a businesswoman and sculptor, invented a secret formula for her own type of artificial stone in the 1770s. At her factory workshop in London she produced material that became famous for being weather-resistant. This ensured that the detail of statues and decorations wouldn't wear away. But Coade didn't just invent her material, she developed applications for it: she employed craftspeople to create designs that could be made from Coade stone. Coade stone can be seen on many historic buildings, including Buckingham Palace.

The statues and decorations on many historic buildings are made from Coade stone; this lion is at Mote Park, Roscommon

Eleanor Coade 1770s

18. Orbital rocket

The rockets that keep satellites in their orbits around the earth

Satellites are small, unmanned spaceships that move around the Earth, which is called 'orbiting'. The moon is a natural satellite. By using satellites we can communicate with people on the other side of the world with phones, TV or the internet. Satellites must be kept on a precise path to keep them in orbit – otherwise they could crash back down to Earth.

It used to be to difficult to steer satellites and keep them on their orbital paths. This was done by controlling a number of small rockets that fired in different directions around the satellite, pushing it this way or the other. In 1967 engineer Yvonne Brill invented a smaller and lighter type of orbital rocket, called a hydrazine thruster, which created a much simpler way of controlling satellites from the ground. Instead of using a number of rockets, her system had a single thruster that fired in different directions.

Brill's innovative invention was first used in 1983. It became the standard way that satellites are steered in space today, earning Brill accolades as a pioneer in space technology.

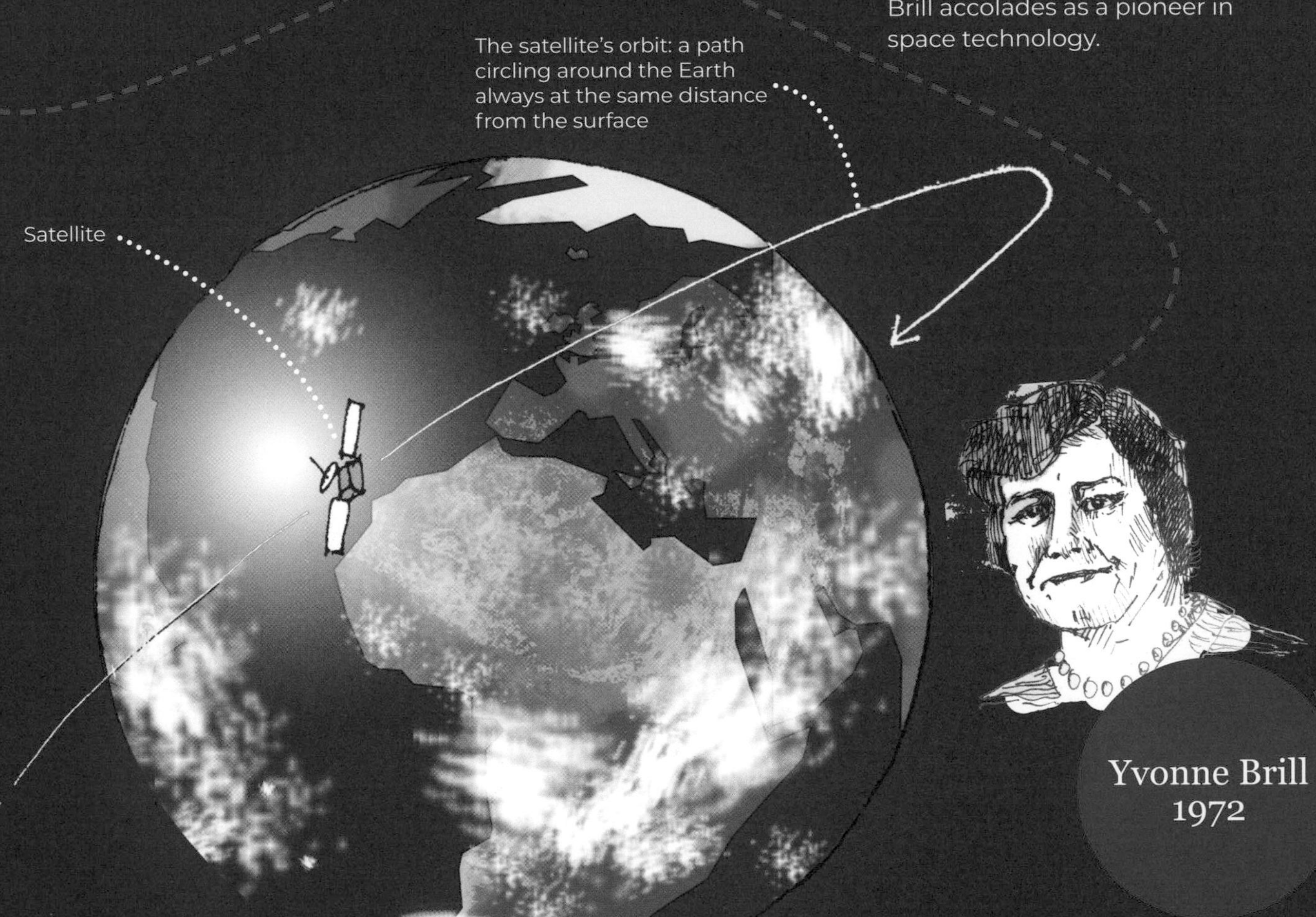

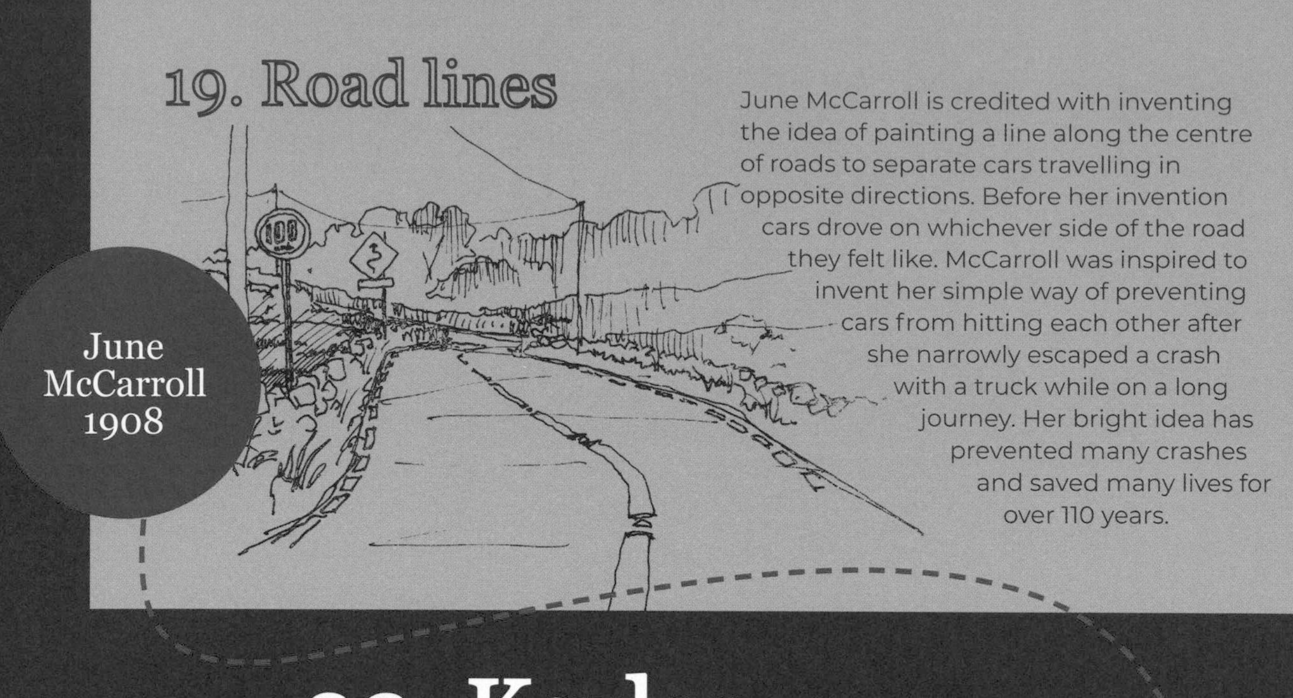

19. Road lines

June McCarroll 1908

June McCarroll is credited with inventing the idea of painting a line along the centre of roads to separate cars travelling in opposite directions. Before her invention cars drove on whichever side of the road they felt like. McCarroll was inspired to invent her simple way of preventing cars from hitting each other after she narrowly escaped a crash with a truck while on a long journey. Her bright idea has prevented many crashes and saved many lives for over 110 years.

20. Kevlar

The lightweight material stronger than steel

Some inventions find applications that are completely different from what they were originally intended to do. One of these is Kevlar, an amazing type of fabric that is five times stronger than steel.

Scientist Stephanie Kwolek invented Kevlar while she was working with her team to develop a new material for making car tyres. They were hoping to create something that would be lighter and stronger than rubber so that cars would weigh less and use less petrol. The material she discovered forms very strong fibres that can be woven into layers to make a hard, protective shell.

Many applications were found for the new material: from bicycle tyres to tennis rackets and in industrial equipment from aircraft to spacecraft parts. It is probably best known as the life-protecting material used to make helmets, safety gloves and body armour.

Stephanie Kwolek 1960

Modern army helmets are now made with Kevlar rather than steel as it is lighter and gives more protection.

Kevlar body armour

21. Using the sun's power

Mária Telkes
1956

Solar power means collecting light energy from the sun and converting it into usable electrical and heat energy. Engineer Mária Telkes pioneered the use of solar power for different applications. One of her most important inventions was a solar-powered device for turning salt water into fresh water. Her invention was used by the US Navy during the Second World War, keeping airmen and sailors in lifeboats alive until they were rescued. Telkes worked on many different projects, including designing solar stoves for use in hotter climates without electricity and a house powered by solar electricity.

The house Telkes designed which was powered by solar electricity

22. Multiplane camera

Lotte Reiniger
1923

How scenes are created for animated films

Charlotte Reiniger was another film-maker – an early pioneer of animated movies. She made beautiful animations with characters that looked like shadow puppets. She made the first ever feature-length animated film: *The Adventures of Prince Achmed* in 1926.

Reiniger wanted a way to create the appearance of depth in her films. She invented her own special device called a multiplane camera. The camera was able to film several pictures made on plates of glass in front of each other at the same time. When filmed in this way the pictures combine together to make the scene. Later other animators copied her techniques.

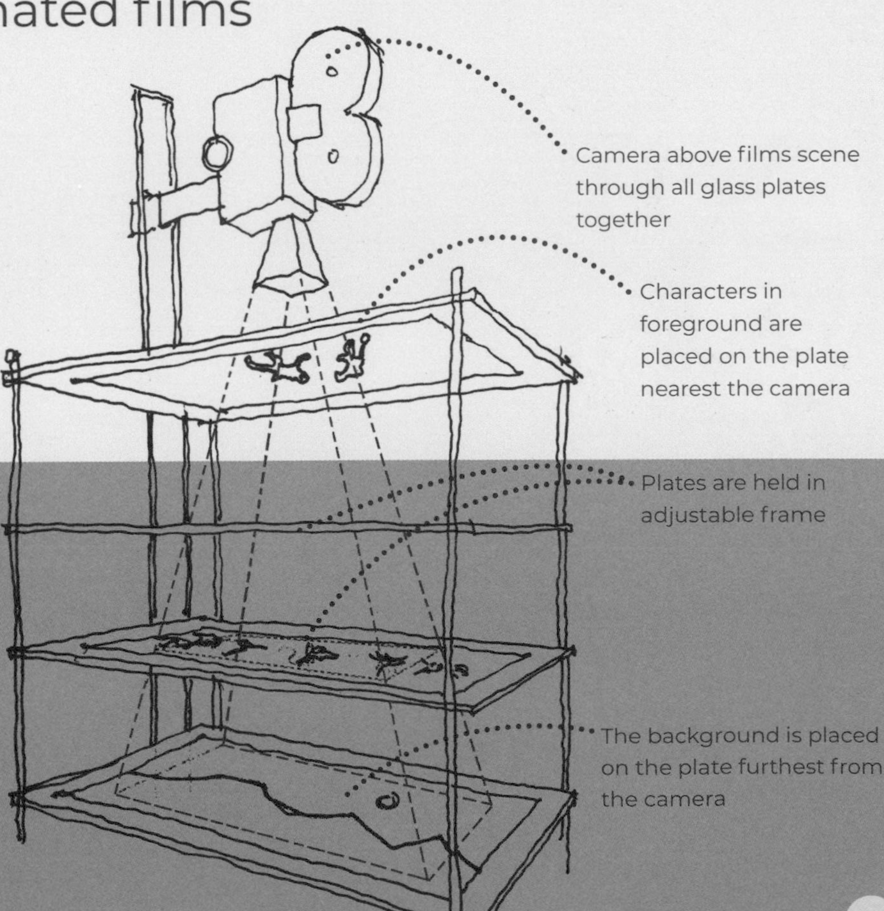

23. Pulsars

The powerful stars that send pulses of energy across the galaxy

Spinning star at the centre

A pulsar (short for pulsating star) is a type of star that rotates and sends out incredibly powerful pulses of energy. They are located far away in outer space, as far as humans can see. Even though pulsars can be millions of light years away they can be detected with instruments on Earth. Astronomer Jocelyn Bell Burnell discovered the first known pulsar in 1967.

Her discovery of pulsars provided valuable information that changed our understanding of how the universe works. Since then people have found interesting uses for the very regular pulses we receive on Earth. They are so precisely regular that they can be used to set clocks very, very accurately. They have even been used to measure the very slow movement of the continents on Earth through continental drift.

Bell Burnell was working as part of a team when she made her discovery. When the team was awarded the Nobel Prize, her name was left out. However, she went on to win recognition for her discovery which transformed astronomy.

24. Life raft

Essential life-saving equipment

Life rafts are essential safety equipment for ships, enabling people to be kept safe on water until they are rescued in case of a disaster. Inventor and engineer Maria Beasley observed many dangers in the existing life rafts used on ships. They floated but didn't provide safety from many dangers of shipwrecks, such as lack of life rafts (due to size and storage space), fire (a big risk on wooden ships), falling overboard or starving while awaiting rescue.

Beasley's ingenious new foldable raft was made of fire-proof material. It had safety rails, to stop people falling overboard, and storage spaces for supplies to keep people alive.

Maria Beasley
1880

25. Laser cataract treatment

A medical device for eye surgery

Patricia Bath
1986

Coloured iris surrounding the lens

With cataracts the lens of the eye becomes clouded and removal is a delicate operation

A cataract is a serious eye condition where the lens becomes clouded, causing loss of sight.

Cataracts need to be treated very carefully in order to restore a patient's vision. The treatment involves: breaking up and removing the damaged lens and inserting a new artificial lens. This procedure requires great care in order to avoid damaging surrounding tissue.

In 1986 scientist Patricia Bath invented a device that uses a laser to break up and remove the cataract: the laserphaco probe. The laserphaco probe is faster and more precise, reducing the risk of damaging tissue.

Bath's career has been dedicated to curing blindness. One of her many achievements was finding a way to help those who might not know they have cataracts to get treatment by creating a system of trained volunteers to test people in their communities.

Inventing tips

Inventing things might seem complicated, but here are some simple steps to help you get started...

Step 1: Be curious

For the inventors in this book, curiosity was the start of their process

HAS ANYONE DONE IT BEFORE?
Find out what already exists. Perhaps there is an already existing invention you could improve on. If someone has already created an identical invention then it might be protected by patent.

WHO IS IT FOR?
Thinking about who might use an invention will help you understand what it needs to do. It's also a good way to think about who might want to buy an invention.

IS THERE A PROBLEM THAT AN INVENTION COULD FIX?
The most important question to ask about an invention is whether it is needed. Can you see a problem that could be solved with an invention? Or an opportunity for a useful new device or system?

Curiosity is very important for inventing and discovering. Noticing problems or opportunities for inventions can start you off on the trail to a new invention. Knight (no. 33) started her inventing career after seeing a factory accident as a child and wondering how to make the machines safer.

Step 2: Pick an idea, set a challenge

This is where you decide what you are going to work on

By now you might be seeing lots of problems or opportunities that could use a good invention. So it's time to pick one to focus on. Set yourself a challenge to invent something that solves the problem or makes the most of the opportunity you have spotted.

Coston (no. 49) focused on an idea to create useful signal flares for ships. Her search for a solution for this idea guided her through ten years of working with chemists to develop her successful invention.

Step 3: Try things out

Think of as many answers to your challenge as you can and test them to see what works

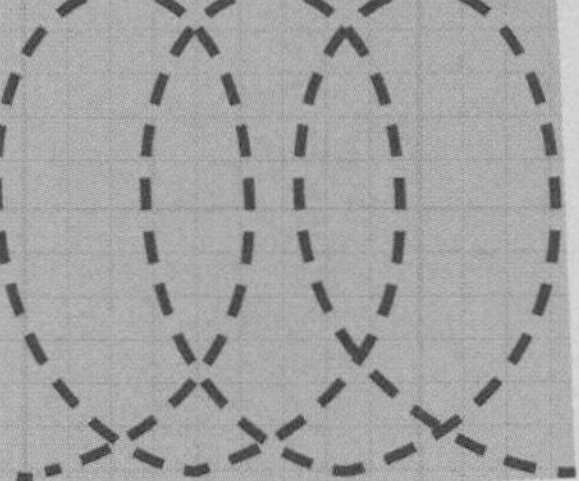

PROTOTYPING
A prototype is a sample version of an invention made early on to help test and examine it. Prototypes also help inventors work out how to make the final product.

USER TESTING
The people who will actually use an invention are often good judges of how well it will work. Asking users to try it out is a really good way of testing, often used for new products with shoppers.

TRIAL TESTING
New medicines are tested using trials. Volunteers take the medicine in a very controlled way and then scientists carefully measure how well it works.

BETA VERSIONS
When writing computer code, programmers often use 'beta versions'. This means starting to use the software in an unfinished form and then discovering and fixing any problems while using it.

MODELLING
A simplified or mini version of the design. Architects often make miniature models of building designs. It's better to discover a problem with a model than a finished building!

Designing and inventing is a process of doing something, checking how well it meets the challenge, improving it and testing again. Benz (no. 39) took the first motor car on its first long distance journey in order to test how well it worked. She and her husband learned a lot from her trial test.

Step 4: Make it!

Use what you learned from your experiments to refine and finish your invention and make it for real

Coade (no. 17) turned her artificial stone invention into a thriving business, building her factory and hiring designers and crafts people.

26. Monopoly

A board game with a message

In 1904 Elizabeth Magie patented 'The Landlord's Game', an ingenious game designed to teach a message. But the game went on to become famous as something quite different.

Since the 1870s Magie campaigned for a fairer society. She believed that one of the main causes of inequality was a monopoly: an unfair property system where a landowner could control most of the property in an area and become extremely powerful.

The game had two sets of rules: the 'Prosperity' rules and the 'Monopolists' rules. Under Prosperity rules every time a player buys a property all players receive some income. However, under the Monopolists rules, when players own property any other players landing on their properties must pay them rent until all but one player go bankrupt.

In 1935 a large games company bought Magie's patents. In 1936 they relaunched the game as 'Monopoly' – with the Monopolists rules only. It became a huge success. Players enjoyed the game without being aware of Magie's message. Magie received no credit and very little money for her invention.

A 1939 edition of Magie's game

Monopoly board

Elizabeth Magie 1904

27. London map

Phyllis Pearsall
1935

Painter and writer Phyllis Pearsall created a design that Londoners have used for many years to find their way around their sprawling city. London is the largest city in Europe with thousands of streets, but Pearsall saw that there were no accurate maps of it, so decided to make her own. She walked around the city, covering up to 23,000 streets and spending up to 18 hours a day noting street names and house numbers. Her map was hugely successful. Not many people know that Pearsall's maps contain fictitious 'trap streets' hidden within them to catch out people who copied her maps without permission!

28. Ichthyosauraus

Understanding the world of the dinosaurs through fossils

Mary Anning is famous for discovering many types of dinosaur including the Ichthyosaurus, which she discovered in 1811. Anning grew up near the beaches of southern England where there were many fossils hidden in the stony cliffs. She made her living from selling fossils. She wasn't a trained scientist but she made many great discoveries and became a well-known expert. One of her most famous discoveries was the first complete skeleton of an ichthyosaurus, which she was able to identify because of her great knowledge. However, Anning wasn't wealthy and didn't make much money from her discoveries. Sometimes trained scientists took the credit for the knowledge gained from her discoveries.

Anning remains famous today because of the tongue-twister written about her: 'She sells sea shells on the sea shore'.

Mary Anning
1811

29. Radioactivity

The powerful energy at the heart of atoms

Most types of atoms are stable and do not change, but certain types of atom, or elements, are unstable and break down over time, emitting energy in the process. This type of energy is called radioactivity. Marie Curie, one of the most well-known scientists of the twentieth century, was the first person to identify, name and understand radioactivity.

This is the symbol used to warn people to be careful of dangerous radiation nearby

Marie Curie
1898

Curie did many experiments to understand how radioactivity worked and discovered new radioactive elements: thorium and polonium, which she named after her native Poland.

Radioactive material is powerful and can be dangerous and needs to be treated carefully. Sadly Curie died of an illness caused by the radiation she was exposed to. Her papers are still too radioactive to handle without protection!

Curie was the first person to win the Nobel Prize twice for her extensive discoveries. She is also the only person to ever win it in two different fields: chemistry and physics.

Curie's daughter, Irène Joliot-Curie, also went on to become a famous scientist. She was awarded the Nobel Prize for her discovery of a method of creating radioactive materials in 1935.

30. Correction fluid

Correction fluid, the fast-drying white liquid that can be used to correct mistakes when writing with pen or typewriters, was invented by Bette Nesmith in 1956. She had worked for a long time in offices. She invented it as her own way of correcting typing mistakes so that a page did not have to be re-typed if there was an error. When other people saw how useful it was, Nesmith developed it into a product. She set up her own company to produce it. It is now used in schools, colleges and offices worldwide.

Bette Nesmith
1956

31. Windscreen wipers

The device to help drivers see and drive safely

Vehicles need windscreen wipers so the driver can see and drive safely in bad weather. Mary Anderson invented the first car window cleaning device in 1903. On a trip to New York City she thought it was strange how tram drivers had to keep getting out to clear snow off the windows. So she designed a device that had a swinging arm with a rubber blade that the driver could move from inside the tram using a lever: the first windscreen wiper.

In 1917 another inventor, Charlotte Bridgwood, improved on it, patenting her design for the first electrically operated windscreen wipers. Bridgwood's daughter, Florence Lawrence, also invented an indispensable safety device. She designed the first indicator signals to show if a car is turning or braking. Now every vehicle needs to have these safety devices by law.

Mary Anderson 1903

32. Y-chromosome

The DNA strand that decides if a person is male or female

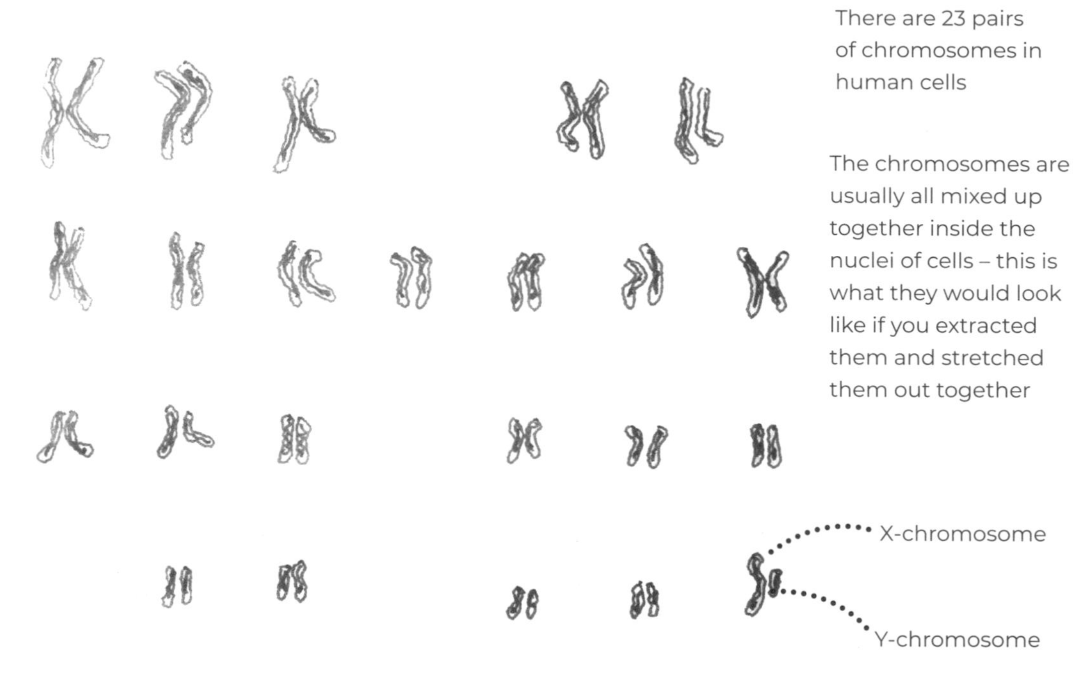

There are 23 pairs of chromosomes in human cells

The chromosomes are usually all mixed up together inside the nuclei of cells – this is what they would look like if you extracted them and stretched them out together

The strands of DNA inside every cell of our bodies takes the form of microscopic strings called chromosomes. Chromosomes are grouped in pairs and contain thousands of genes that are what determine a person's physical traits, such as the colour of their eyes or hair.

Nettie Stevens was a pioneering biologist who studied beetle larvae in order to understand chromosomes. From her observations she was able to work out that chromosomes determined whether an organism was male or female. She discovered that only one pair determined gender: X-chromosomes.

She identified that often one of these chromosomes was actually a different type: the Y-chromosome. Stevens found that in female larvae these pairs were made up of two X-chromosomes (XX). In male larva they were made of a X-chromosome and a Y-chromosome (XY).

From this she worked out that the Y-chromosome is what makes an organism male, and that males determine the gender of their off-spring. This is known to be true for nearly all creatures.

Nettie Steven
1905

33. Paper bag

The shopping bag design used all over the world

Margaret E. Knight invented many things. She came up with her first invention at the age of twelve when she invented a safety device for looms after witnessing an injury in the textile mill where she worked.

In 1868 she designed a machine to fold and glue paper bags with a rectangular base to give them their shape and strength. Her machine took a continuous tube of paper from a roll, then grooved, folded and cut it to form the bags. This was so successful that shopping bags throughout the world today are based on her design. While she was developing the invention, another inventor tried to steal her idea, but Knight successfully won a court case to prove she thought of it first.

Margaret E. Knight 1868

34. Disposable nappies

Sometimes different designers come up with the same idea separately. Marion Donovan and Valerie Hunter Gordon both patented disposable nappies within a year of each other. Both designs featured a waterproof outer cover (to prevent nappy rash) and an absorbent pad inside that is disposed of after use (reducing the work of washing). They were were inspired by solving problems they had experience of: babies getting rashes and a lot of washing. The world of nappies has since moved on. Today we need to make them more sustainable, because disposable nappies are not good for the environment. However, the ingenuity of their idea endures.

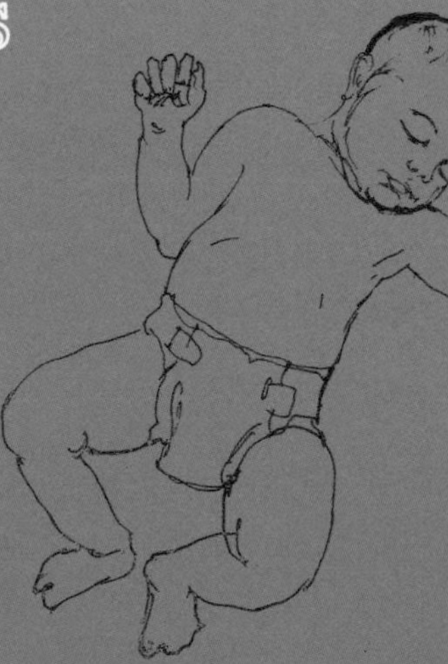

Marion Donovan 1946 / Valerie Hunter Gordon 1947

This picture shows Hamilton standing beside a print-out of the computer programme for the Apollo missions!

35. Coding to the moon

Designing the software that enabled the moon landings

The Apollo missions were a series of space expeditions by American astronauts to the moon using rockets. The Apollo 11 mission was the expedition where a human first landed on the moon. Mathematician and physicist Margaret Hamilton led the team that created the software that ran the computers on the Apollo rockets.

Hamilton's special expertise was in how to keep computer systems working if there was a problem. This was crucial during the first landing, when an error occurred in one of the systems. Hamilton's software design enabled the computers to keep running and prevented the mission from being aborted, ensuring that the astronauts were able to make the first landing on the moon.

Hamilton was a pioneer in her field, and invented the term 'software engineering' to describe the process of making and managing computer programs.

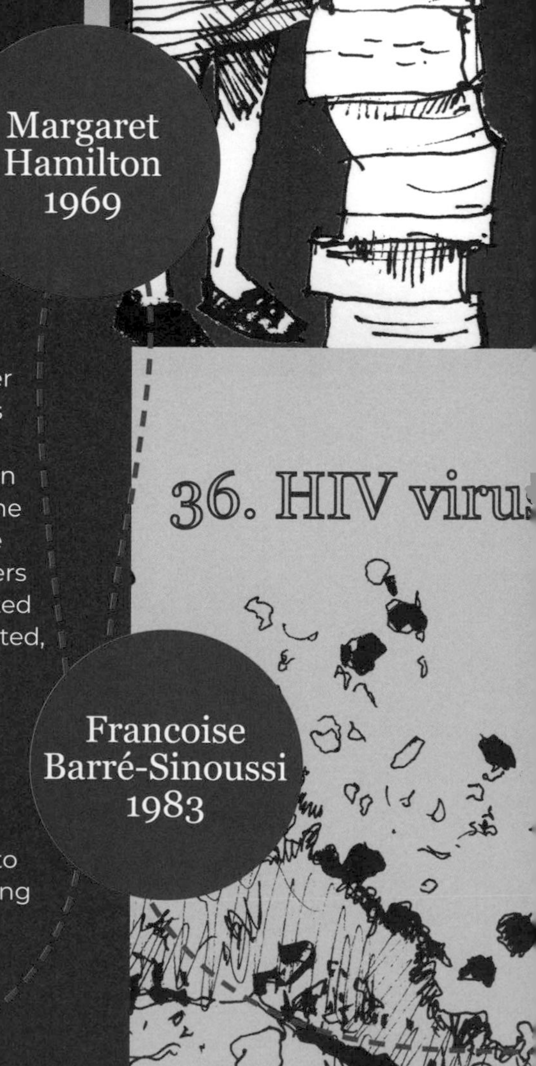

37. The lives of jellyfish

Untangling the mysteries of ancient sea creatures

Jellyfish have existed in our seas for over 500 million years and are one of the oldest types of sea creature on Earth.

Jellyfish have two forms during their life cycle. The young hydra form clings to rocks while the fully grown medusa form floats with long tentacles. For a long time the relationship between these two forms was unknown.

From her home on Valentia Island, off the southwest coast of Ireland, an amateur scientist called Maude Delap was determined to solve the mystery.

By making her own aquariums, she succeeded in breeding captive jellyfish. By working out the life cycles of the Lion's Mane jellyfish and Compass jellyfish, she unravelled the mystery of their hydra and medusa forms, learning which hydra belonged with which medusa.

Delap was a gifted scientist, but despite being honoured for her work and offered a position abroad, her father didn't permit her to move away without being married first. She remained on Valentia and lived for many years, well known for her knowledge and discoveries.

When the disease AIDS (acquired immune deficiency syndrome) became more widespread during the 1980s, doctors and scientists didn't know what was causing it.

Scientist Francoise Barré-Sinoussi joined the search for the cause. She was an expert in the study of retroviruses. Retroviruses are a type of virus that spread in the body by inserting their genetic material into the DNA of a sufferer's cells. She studied patients who were showing early signs of having AIDS. She found evidence that a new kind of retrovirus was attacking the patients' lymphocytes – a type of blood cell. These cells are crucial for fighting diseases. She discovered that the retrovirus attack on these cells was the cause of AIDS.

Barré-Sinoussi's discovery of how the disease worked made it possible to create treatments. She was awarded the Nobel Prize along with the other scientists she worked with for her discovery.

Maude Delap
1902

Ways of discovering

The women in this book who made discoveries worked in different ways. But there are some scientific techniques that were common to all of them. These are great ways to explore topics that you are interested in yourself.

TESTING

Scientists often start their research with a hypothesis about how something works. They test the hypothesis through experiments. The results are measured and recorded and the process is repeated, adding new information.

Judge, Hickey and Healy-Thow (no. 16) discovered how rhizobium bacteria speeds up seed growth by experimenting to measure the germination of 11,000 seeds.

EXPLORING AND OBSERVING

Exploring involves examining things closely to understand how they work. Using microscopes to see tiny organisms is an example of exploration. Examining things over time to understand how they change is also important.

Crowfoot Hodgkin (no. 45) used her X-ray crystallography technology to explore and observe molecules in a way that wasn't possible before. This helped her discover the structure of Vitamin B12.

IDENTIFYING

Identifying and classifying is important to understand how the world works. When scientists examine new or materials, they work out how they might relate to other things that are already known.

With pattern seeking, scientists examine many things or events over time to find connections. They use the inforrmation to make models of how they think things work. Improvements in the power of computers makes it possible to examine data to find patterns.

FINDING PATTERNS

Tharp (no. 2) discovered the Mid-Atlantic Ridge by finding patterns in surveys of the ocean floor from mapping expeditions. Her discovery proved that the model of continental drift was correct.

Delap's studies of the Compass and Lion's Mane jellyfish helped scientists understand these species by working out how they related to each other and to other sea creatures.

SCIENTIFIC MODELS

A scientific model is a description, diagram or image used to try and understand how something works when it is difficult to see directly. Models continue to change when new information is discovered.

Meitner's (no. 12) discovery of nuclear fission provided a model for explaining how uranium nuclei break down into smaller elements.

BY ACCIDENT!

Often discoveries are made by chance. Scientists are prepared for the unknown. They are ready to change their ideas to match unexpected results and new facts.

Bell Burnell (no. 23) discovered pulsars by accident while researching other data. At first her team had no idea what caused the strange energy pulses and even considered them alien signals. Later she figured out that the signals came from pulsars.

38. Frankenstein's monster

Mary Shelley
1818

The terror of an invention that gets out of control

Not all inventions are devices. Mary Shelley invented a creature that has remained terrifying 200 years later: Frankenstein's monster. Shelley's character was created during a scary story competition with friends. Doctor Frankenstein makes a living creature out of other body parts and brings it to life. But then his creation gets out of control...

The novel, published in 1818, focuses on the risks of doing experiments that have unknown outcomes. It also highlights the ethics of experimenting with living things, for example, a new living being which becomes a monster.

In the novel the doctor's creation has no name. The famous Frankenstein's monster look shown in this picture was designed for a film adaptation in 1931.

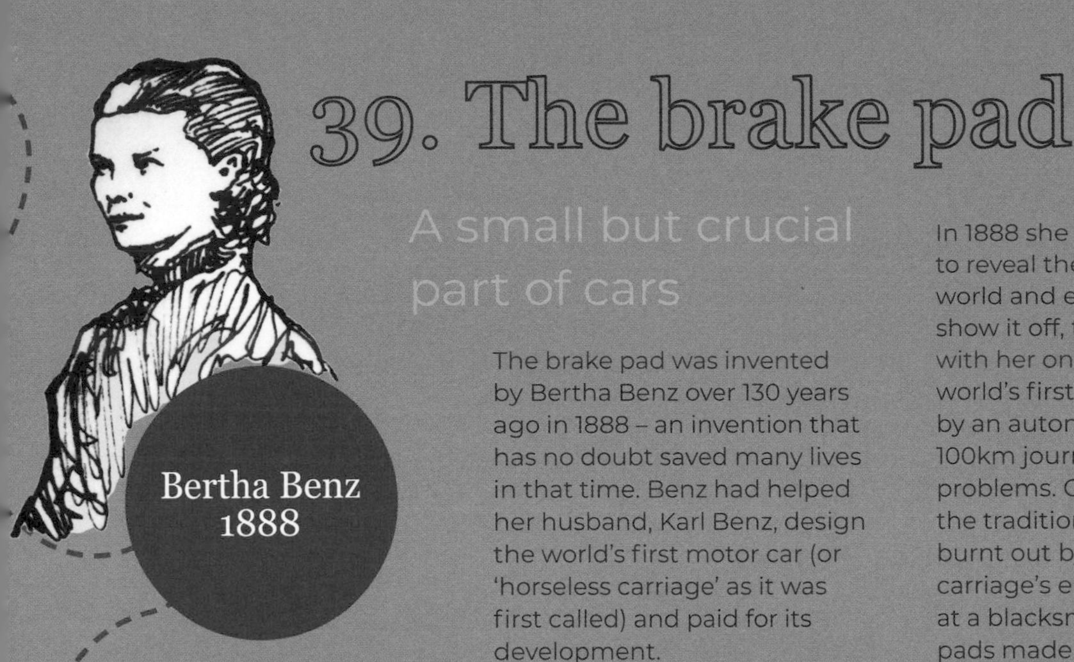

39. The brake pad

A small but crucial part of cars

The brake pad was invented by Bertha Benz over 130 years ago in 1888 – an invention that has no doubt saved many lives in that time. Benz had helped her husband, Karl Benz, design the world's first motor car (or 'horseless carriage' as it was first called) and paid for its development.

In 1888 she decided it was time to reveal the invention to the world and embarked on a trip to show it off, taking her two sons with her on what became the world's first long distance drive by an automobile. While on the 100km journey she fixed several problems. One problem was that the traditional brakes were being burnt out by the power of the carriage's engine. Benz stopped at a blacksmith's and had leather pads made which fixed this – inventing the first brake pads!

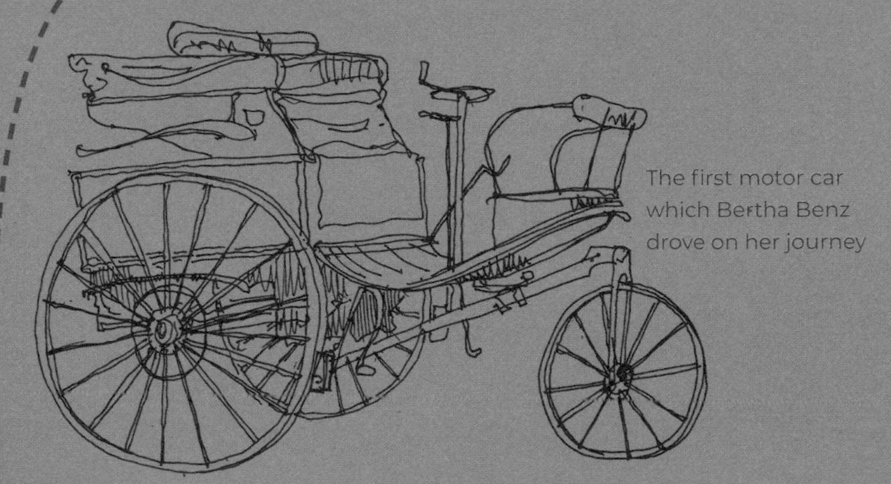

The first motor car which Bertha Benz drove on her journey

Benz demonstrated the need to test models to find out how to improve them, a process called prototyping. From the discoveries made on Benz's trip several changes were made to the car, and the publicity she received for the new invention made the car famous.

40. GM crops

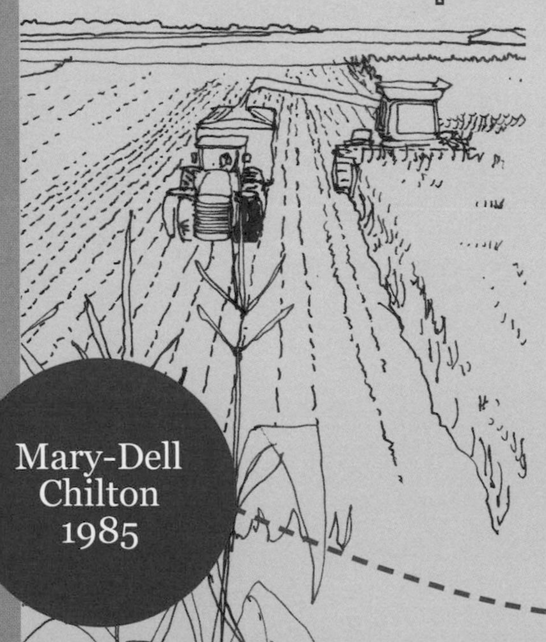

The science of genetically modified crops (or 'GM Crops') is based on discoveries made by scientist Mary-Dell Chilton. Chilton was fascinated with genetics, an area of science opened up by the discovery of the structure of DNA in 1953 (see no. 8).

'Genetically modifying' means changing the DNA inside plant cells to change the characteristics of the plant. For example, a crop's DNA could be changed to make it more resistant to insects so that farmers need to use less insecticides. While GM crops have advantages, some people have worries about creating new living things that don't occur in nature and are concerned that the changed DNA might have unintended side-effects.

41. Villa E-1027

A special house that broke the rules

Modernism was a movement that used new materials and forms to create innovative buildings and objects. E-1027 was one of the first modernist houses. Designed by Irish architect and designer Eileen Gray, it was built in the south of France.

On a steep, rocky site above the Mediterranean Sea, E-1027's flat canopies and terraces made it look more like part of an ocean liner than a house. Gray filled her unique open-plan house with her own distinct furniture designs, often using materials more commonly found in industrial buildings. For example, she was the first person known to use chrome as a material in furniture; this became popular with other modern designers.

E-1027 was a code for the initials of Gray and her partner at the time, Jean Badovici. When it was built it had a big effect on other designers. E-1027 and Gray's furniture continue to influence designers today. The villa is now open to the public.

Eileen Gray
1929

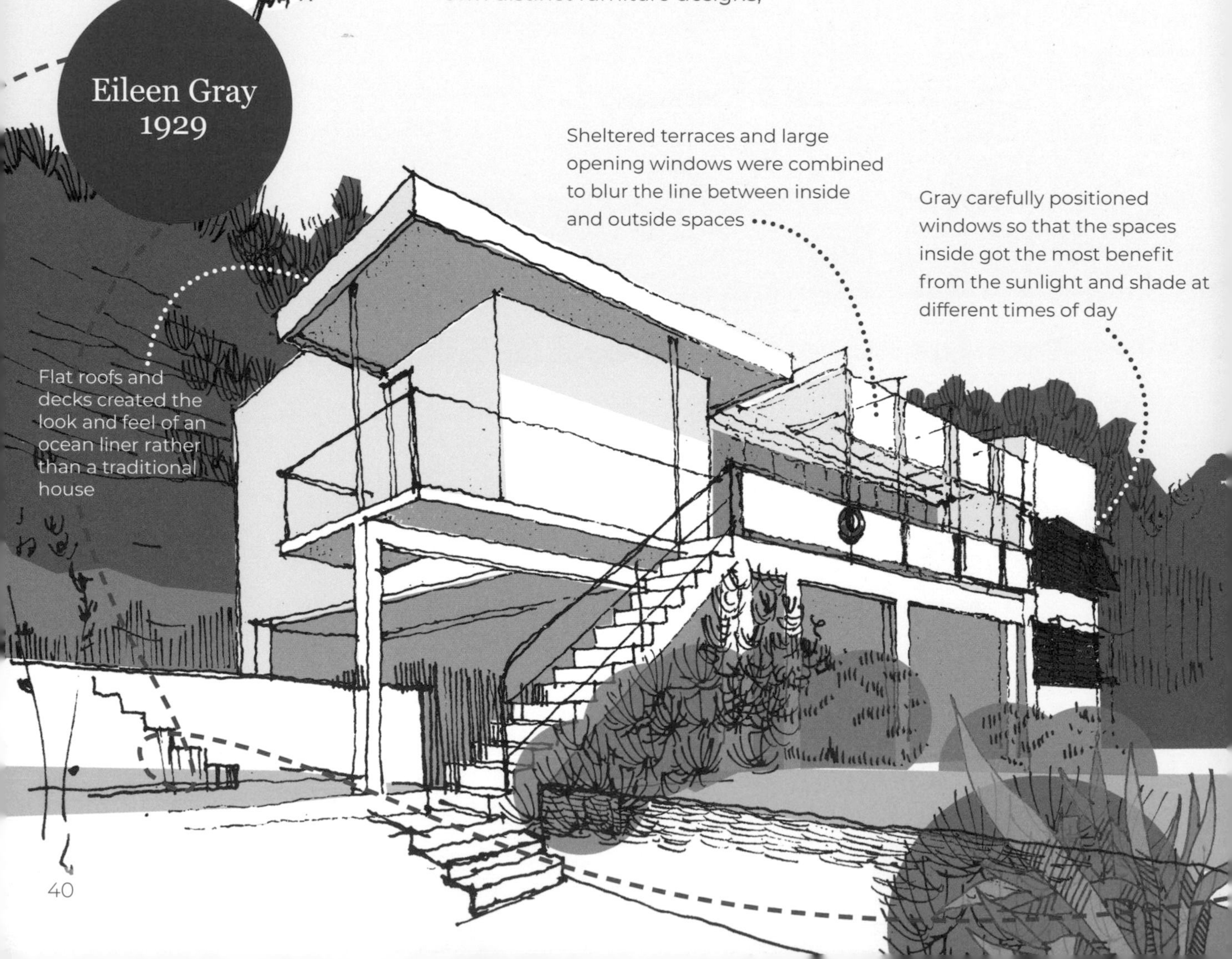

42. Safer wood chopper

Ayla Hutchinson 2016

Another young inventor, Ayla Hutchinson, was thirteen when she witnessed her mother accidentally cut herself with a hatchet while chopping wood. Hutchinson saw that there could be a safer way to chop wood. She first created her invention for a science fair.

Her prototype consisted of an axe blade set inside an iron safety cage. Instead of swinging an axe, the user places a log on top of the blade safely inside the cage and hits it with a hammer, causing it to split on the blade. The design has several advantages. It is easier to use than holding the wood and aiming an axe at it. The blade is kept inside a safety cage instead of at the end of a swinging axe and it is easier for physically impaired people to use than an axe. Called the Kindling Cracker, her invention is now at the centre of a growing business.

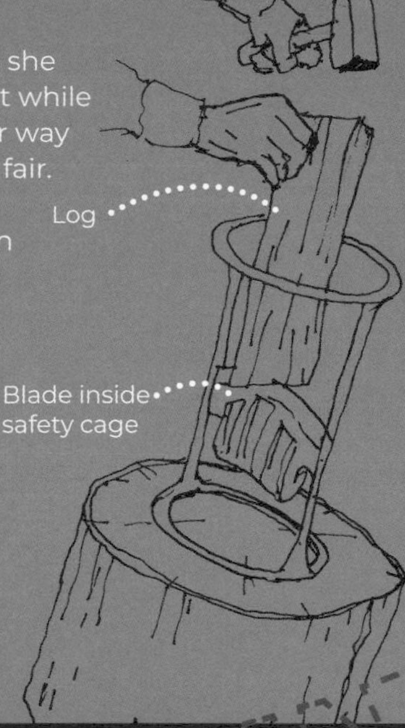

43. Secret signals

Technology for securing communications

During the Second World War, Hollywood film star Hedy Lamarr took on a new role as an inventor. She patented a way to stop enemies from jamming radio-controlled torpedoes. Her invention worked by making the radio frequencies between the torpedo and the controller change frequency in a random way – a method called 'frequency hopping' or 'spread spectrum technology'.

She offered her idea to the US Navy, but the idea was ahead of its time. They rejected her proposal, but they did make use of this technology twenty years later.

Today spread spectrum technologies are used as a way of keeping all of the many millions of phone signals taking place every moment separate from one another.

Hedy Lamarr 1942

44. Science of nursing

Studying disease to reinvent how patients are cared for

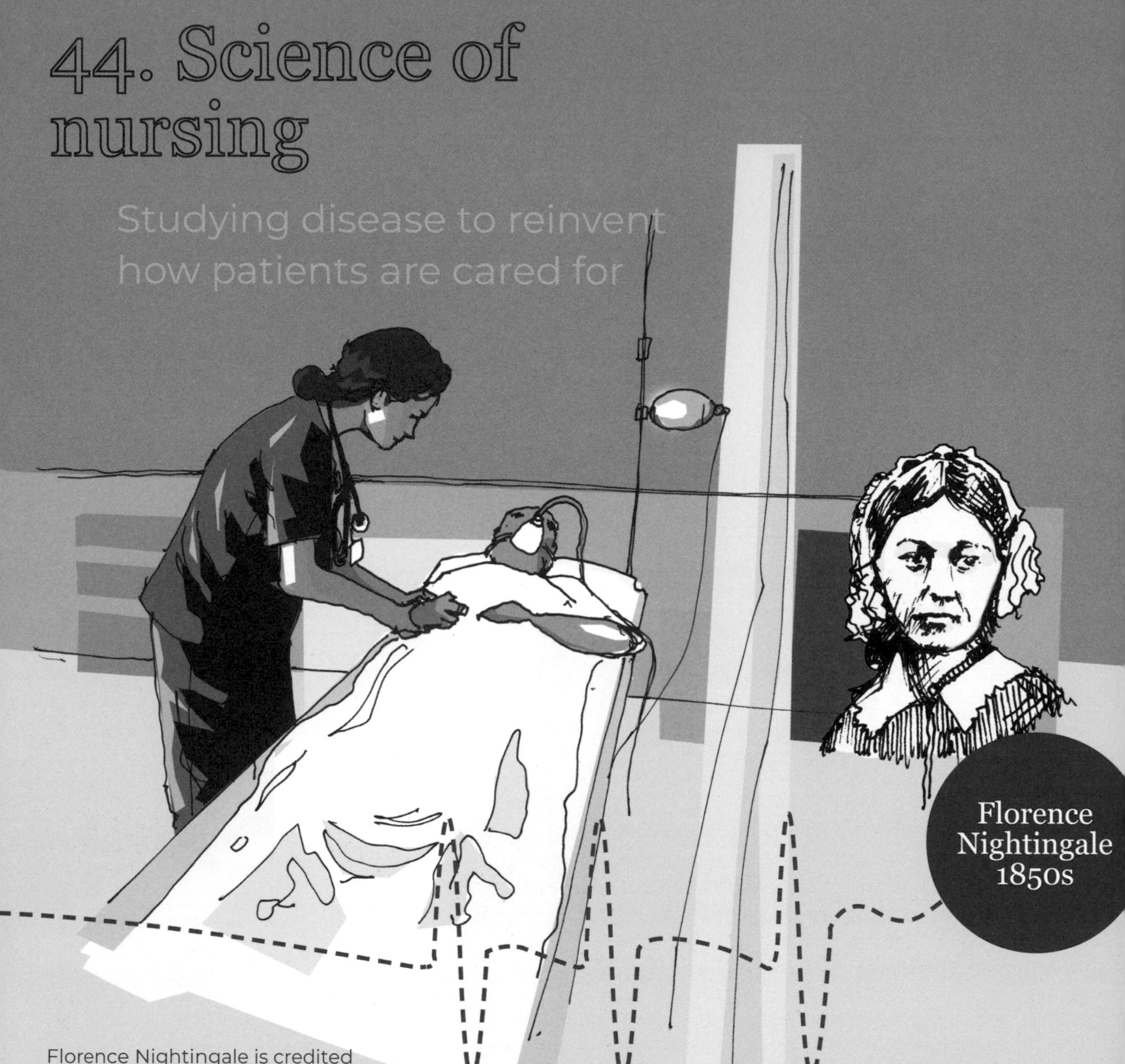

Florence Nightingale is credited with inventing the profession of nursing.

She was a gifted mathematician and put these skills to use analysing information about disease. She was an expert communicator, able to explain her ideas using simple language and infographics that were easy to understand.

Nightingale was committed to improving the lives of people – in particular, wounded soldiers in the Crimean War. She trained and brought a team of nurses to the warzone. She realised that more soldiers were dying from diseases than battle injuries. So she radically changed the design of the hospitals to introduce light and fresh air, reducing the spread of diseases.

Considered a hero, she became legendary as the 'Lady with the Lamp', tending to patients at night. She set up the world's first professional nursing school. While Nightingale might be famous for the caring nursing she did herself, it is her life-saving scientific methods that had the greatest impact and are still relevant today.

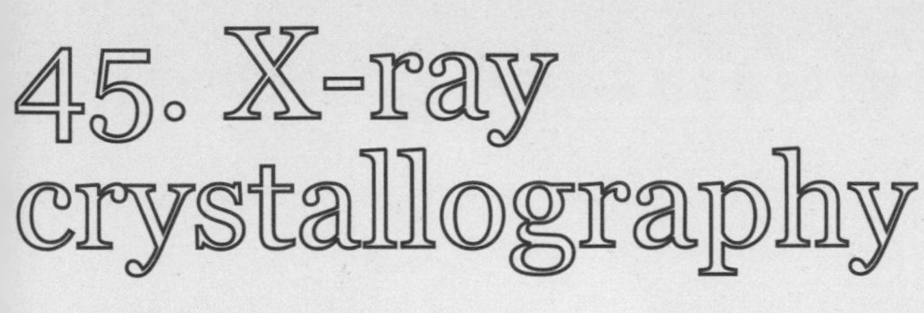

45. X-ray crystallography

Exploring the structure of molecules

A diagram of the structure of the penicillin molecule

Molecules are combinations of atoms that make up the most basic building blocks of a material. The structure of molecules can be seen by examining them with X-rays: a process called X-ray crystallography.

Dorothy Crowfoot Hodgkin pioneered using X-ray crystallography to explore biological molecules.

With her pioneering methods she discovered the molecular structures of Vitamin B12, penicillin and cholesterol. Her discovery of penicillin's structure allowed life-saving antibiotics to be be produced artificially for the first time. Franklin (no. 8) consulted Crowfoot Hodgkin during her research on the structure of DNA. For her discoveries, Crowfoot Hodgkin was awarded the Nobel Prize for chemistry in 1964.

46. Plastic from bananas

Elif Bilgin
2013

Standard plastics produced from oil cause pollution when they are thrown away because they don't break down over time (or bio-degrade) like naturally occurring materials.

Sixteen-year-old Elif Bilgin was shocked at the plastic pollution in the sea near her home in Turkey. She decided to find a new way to make a plastic from naturally occurring material, called a bio-plastic. She experimented using chemicals with banana skins. These are rich in starch that can be used to make plastic. Bilgin found a way to make bio-degradable plastic from the material.

Her sustainable invention could reduce the amount of plastic used for packaging and reduce waste by finding a new use for the food material that would otherwise be thrown away.

47. Earth's inner core

The secrets at the heart of the Earth

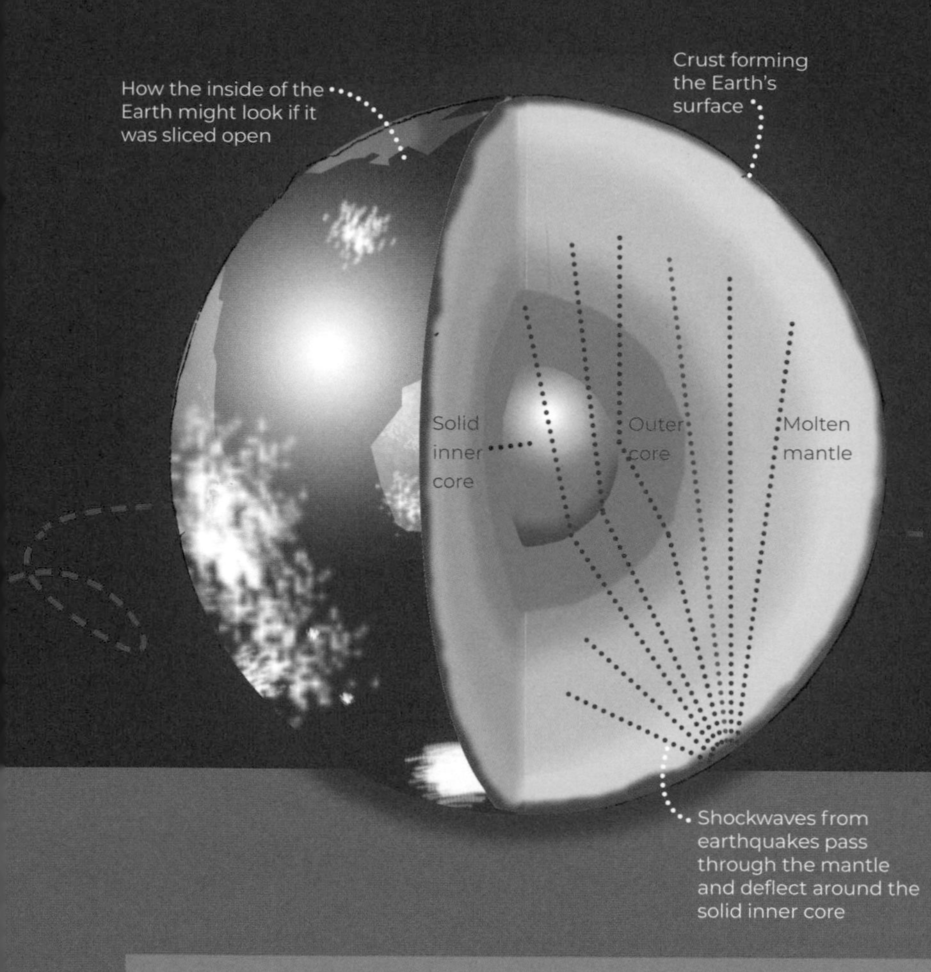

Inge Lehmann 1936

When earthquakes, volcanoes or landslides happen, they send shock waves, or seismic waves, travelling through our planet's crust. Special scientists, called seismologists, measure these waves to learn about the earthquakes and how the Earth's crust of rock and the layers of magma (liquid rock, like lava) are made up. Scientists originally believed that Earth's core was a single molten sphere.

However, while researching an earthquake in New Zealand in 1929, seismologist Inge Lehmann discovered that the seismic waves did not fit the expected pattern. She calculated that this must be caused by a solid core at the centre of the Earth. When her discovery was published in 1936 it changed our understanding of how the inside of the Earth works.

48. Montessori system

Maria Montessori 1907

The Montessori system is a method of learning for young children that has benefitted people all over the world. Invented by Maria Montessori in 1906, this child-centred approach was considered revolutionary at the time. Montessori was a teacher who disagreed with traditional teaching methods where children were just expected to memorise information. Her new system allowed children to learn by doing things themselves. This empowered them by growing their independence, creativity and problem-solving. In Montessori classrooms children are free to move around and learn through being active. Her invention became a huge success and is now used in schools throughout the world.

49. Signal flares

Ships communicating before radios

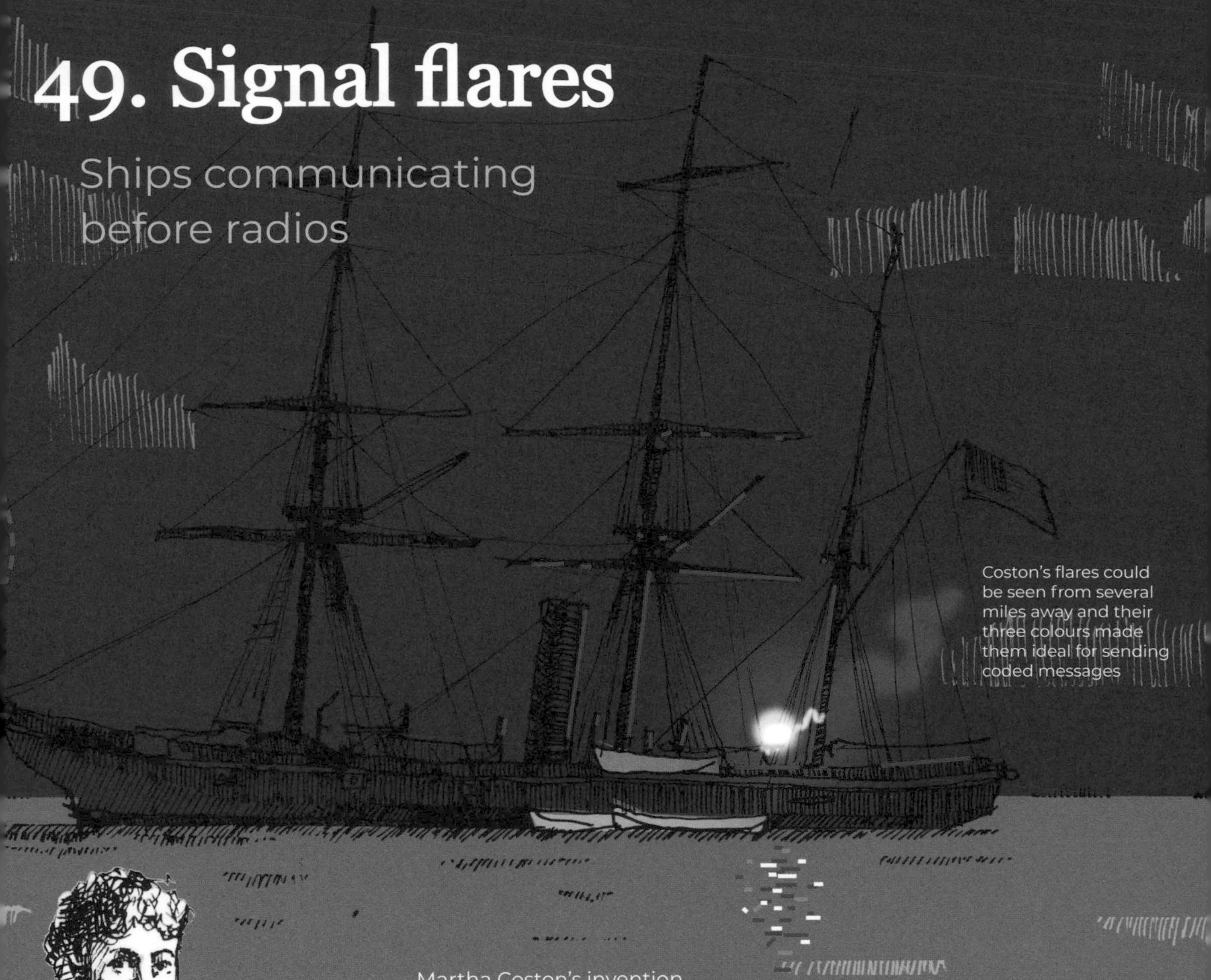

Coston's flares could be seen from several miles away and their three colours made them ideal for sending coded messages

Martha Coston
1859

Martha Coston's invention has proven very important to seafarers for over a century and been crucial in many emergencies. Before radios and satellites existed, ships had to have a way of communicating with each other and with the shore over long distances or at night.

Coston's husband, Benjamin Franklin Coston, had died young in 1848, leaving behind an idea for a signalling flare. Martha Coston set to work and spent ten years turning this idea into a usable invention. She patented the Coston flare in 1859 and the US Navy immediately purchased it. In 1861 the American Civil War broke out and the Navy successfully used Coston flares to communicate between ships during the conflict. Their use quickly spread to ships and rescuers around the world.

Coston invented a code for signalling using different colours so that ships could send messages. Coston flares are still used on ships and lifeboats today to signal to rescuers in emergencies.

50. Hydrometer

The ancient way to measure the density of liquids

The hydrometer is a hollow glass tube with a wider bottom to help it float. A weight at the end keeps it upright

The density car be measured by reading the liquid's level off the measuring scale

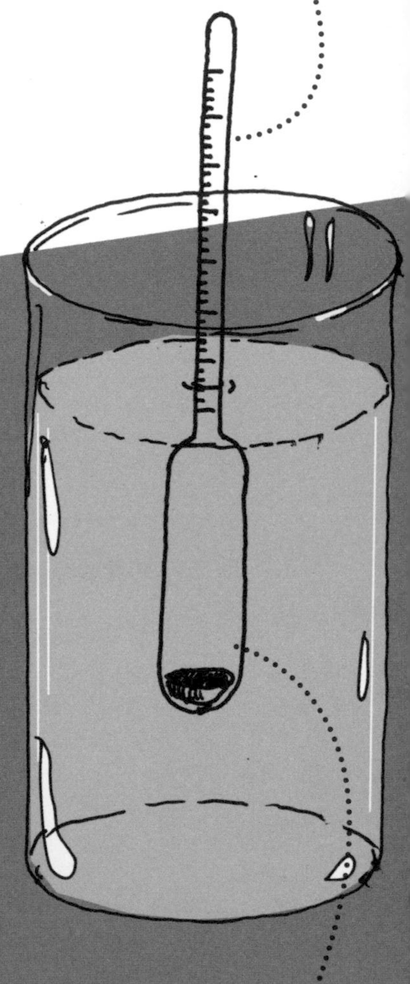

Hypatia of Alexandria c.400

The hydrometer sinks lower in a less dense liquid

It floats higher in a mor dense liquid (e.g. water with sugar dissolved in it would be denser than water alone)

Some inventions were made so long ago that it can be hard to say who's responsible, but the earliest instructions about how to make a device called a hydrometer have been traced to Hypatia, an ancient philosopher and mathematician. She lived in Alexandria in Egypt in approximately 400 CE.

The hydrometer is used to identify the density of a liquid by comparing it to the density of water. It works by pouring a sample of the liquid into a container and floating the hydrometer in the sample.

If the hydrometer sinks low it shows that the liquid is less dense. If it floats higher it shows that the liquid is denser. The measurement of the liquid's density is read off the side of the hydrometer by noting the level of the liquid against the measuring scale. Hydrometers have many uses; for example, measuring the amount of sugar in a drink: the more sugar there is the denser it is.

Looking after your ideas

Why do some people become famous for an invention and not others? This has been a problem throughout history and there have been too many cases where women have not been recognised for their creations. To be given proper credit for your innovation involves telling people that the thinking work you have done (called intellectual property) belongs to you. There are different ways to protect intellectual property and get credit for discoveries or inventions.

For scientific discoveries, the most common way to get credit is by publishing the research in an article in a scientific journal. This publicises what was discovered and when it was found.

For inventions, patents are the normal way to give credit. The really important thing about a patent is that it prevents other people from copying an inventor's idea for a long time so that the inventor has time to turn the invention into a success.

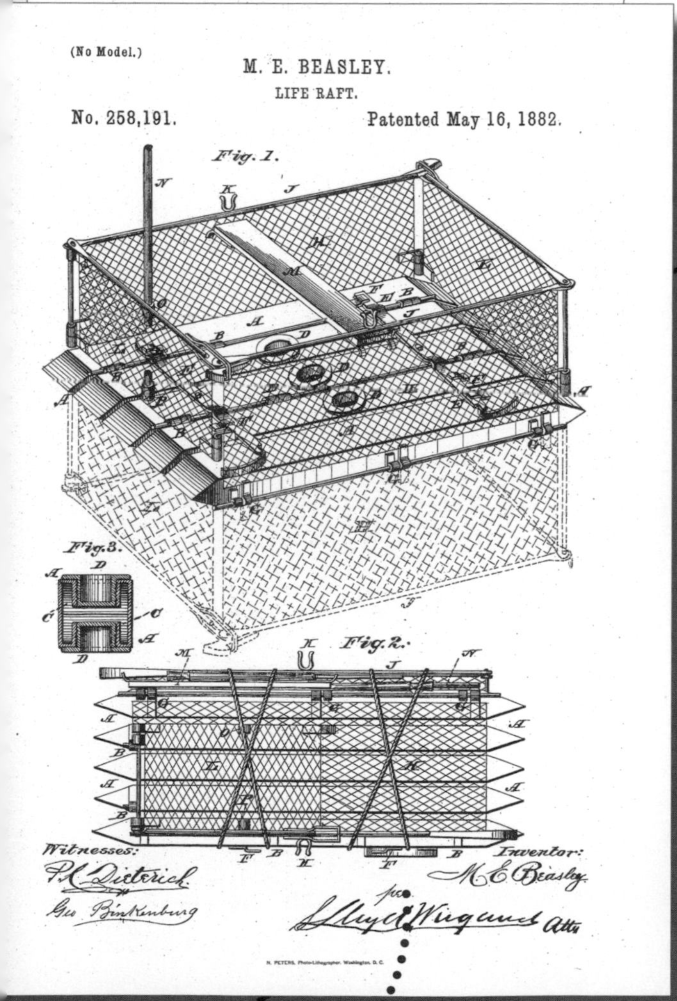

(No Model.)

M. E. BEASLEY.

LIFE RAFT.

No. 258,191. Patented May 16, 1882.

These drawings are part of Beasley's patent for her life raft invention. The patent shows how it works, why it is a new idea and proves when she invented it

For designs or other creations like art, writing or music, people protect their work with copyright, which means that no-one else can make a copy of the work without the designer or artist's permission.

However, credit is not always clear-cut. More than one person can be involved in a discovery or invention. For example, with very advanced science or technology, one person couldn't possibly do all the work. Or inventors might rely on earlier inventions that their work improves.

So, if you are the first person to come up with an invention, see how you can protect your idea. And if you work as part of a team, make sure that everyone who helped is acknowledged and gets credit!

Index

Things

Suspension bridge (4) 10
Syringe (10) 14
Using the sun's power (21) 23
Villa E-1027 (41) 40
Windscreen wipers (31) 31
X-ray crystallography (45) 43
Y-chromosome (32) 32

People

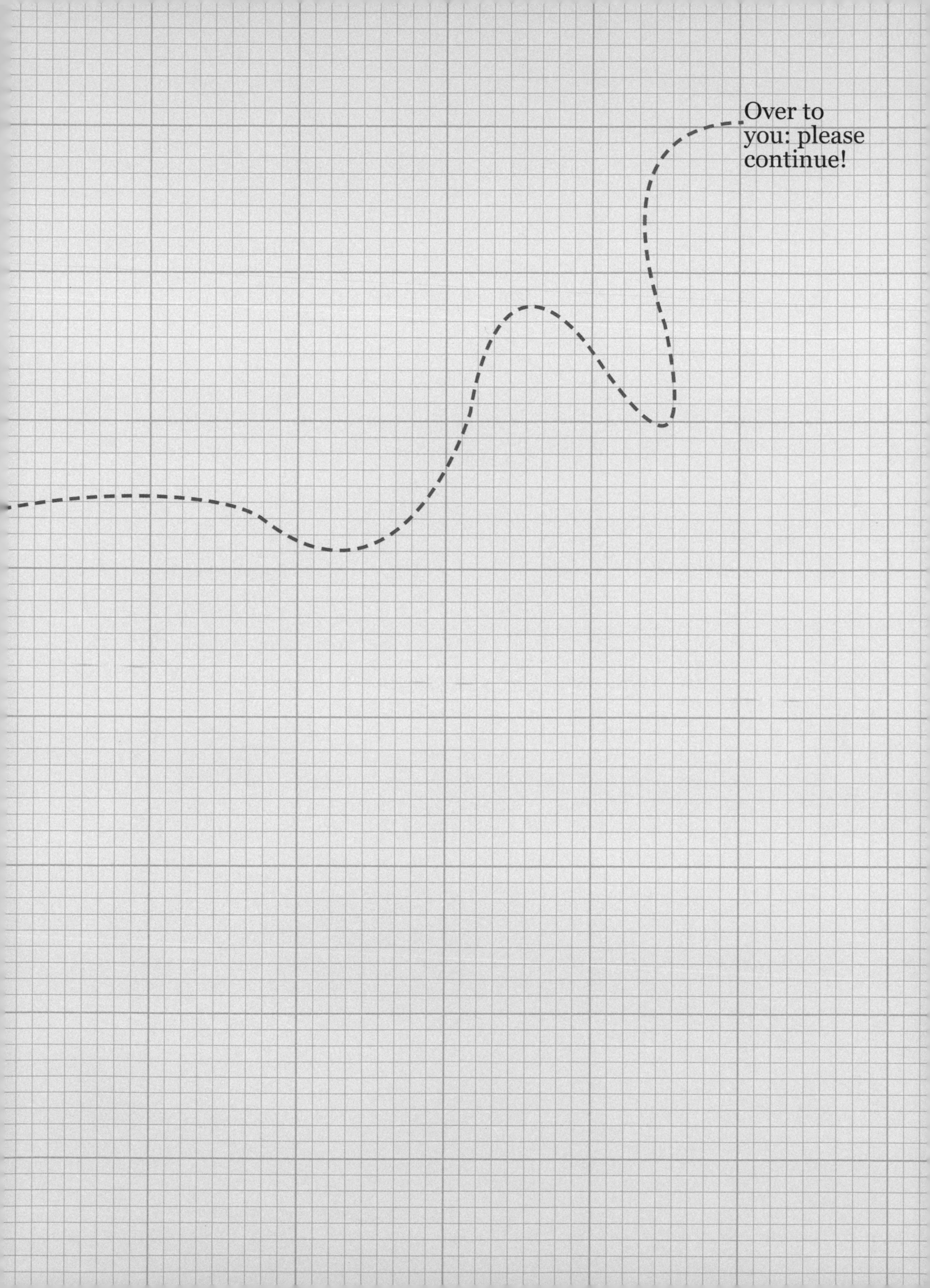
Over to
you: please
continue!